USING COMPUTERS IN PHYSICS

USING COMPUTERS IN PHYSICS

John R. Merrill
The Florida State University

Houghton Mifflin Company • Boston
Atlanta • Dallas • Geneva, Illinois • Hopewell, New Jersey • Palo Alto • London

Printed in the United States of America
Library of Congress Catalog Card Number: 75-25012
ISBN: 0-395-21411-4

contents

foreword

Uses of the Computer

The text material is meant to suggest ways in which the computer can increase your intuitive understanding of physics. The book is to be used in conjunction with your course text. The computer is used in two ways. The first allows you to produce a number of examples of the physics discussed in your text. For example, you can draw more complicated field and potential maps, or intensity patterns from more complicated wave source patterns. The computer allows you to do many more of the things you would have been doing in a course anyway.

The computer also allows you to compare the idealized physical situations of the textbook to more realistic physical situations. Examples are computer calculations of waves in dispersive media, or calculations of interference patterns from radio and television antenna patterns.

Uses of Textbooks

You are expected to learn the basic derivations and relationships from your textbook. Only a brief statement of the equations and laws being used will be given here. (The equations are cited so that you can see which chapters of your text cover the same material.) The strategy of a computer approach will be outlined, and computer programs will be introduced. Figures drawn by a computer are included, and chapters include some suggested problems.

Programs

The programs presented are not as efficient as possible. Instead, they are written to make the method clear. You can often write substantially better programs in terms of calculational speed and required storage space. In some cases you will need to rewrite the programs to match the specifications of a particular computer. All the programs presented have been executed on the Dartmouth H-635 computer. Every effort has been made to use only the simplest computer instructions. Since languages differ from computer to computer, you should consult the manuals for your computer system.

Units

The material makes use of "scaled units," which rescale lengths, time, and so on, so that the equations have a particularly simple form. Scaled units are used heavily by theoretical physicists in particular, but they have an extra advantage when we are using computers, since the numbers being used in calculations remain near 1 (rather than becoming too small or too large for the computer). For example, in quantum mechanics the basic equation is

$$-\frac{\hbar^2}{2m}\frac{d^2\psi}{dx^2} + V(x)\psi = E\psi$$

By scaling the units (so that lengths and energies are measured in atomic-sized units) the equation becomes

$$-\frac{1}{2}\frac{d^2\psi}{dx^2} + V(x)\psi = E\psi$$

Very small numbers such as the constants $\hbar$ and m are removed; the physics is unchanged. Watch for scaled units and be sure that you measure lengths and so forth in proper units when you compare your calculations to experiment.

Flow Charts

Simplified flow charts (block diagrams) are given with a number of the programs in this book. The flow charts are meant to illustrate the strategy implemented by the program and not the program itself. Many possible programs are correct implementations of any strategy, so the important thing is the strategy of the calculation. Those of you who are used to full flow charts of programs, with all their internal loops fully disclosed, for example, may find these block diagrams need supplementing. However, more often than not full flow charts obscure the strategy rather than making it clearer by a graphical presentation.

BASIC and FORTRAN Instructions Every effort has been made to use only commonly available BASIC and FORTRAN IV instructions. All the programs have been executed on the Dartmouth Honeywell computer; many have also been executed on a number of other computer systems. Nonetheless, it is very possible that small changes will be necessary to meet the particular demands of particular computer installations. Consult the language manuals for your computer.

BASIC

Arithmetic operations
+, −, *, /, ↑

Often used instructions
LET
PRINT
INPUT
READ – DATA
END
IF – THEN
GOTO
DIM
FOR – NEXT

Less used instructions
DEF
MAT A=B
MAT A=B*C
MAT A=(scalar)*B
MAT INV

Often used functions
SIN
COS
ABS
SQR

Less used functions
RND
INT
TAN
ATN

FORTRAN

Arithmetic operations
+, −, *, /, **

Often used instructions
ANUM= .01 (assignment)
PRINT
WRITE
READ
DATA
FORMAT
STOP
END
GOTO
IF ( ) #,#,# (arithmetic if)
DO
CONTINUE
DIMENSION

Less used instructions
IF (relation) (logical if)
FUNC(X) (statement function)
COMPLEX, REAL

Often used functions
SQRT
SIN
COS
ABS

Less used functions
ATAN
COMPLX
FLOAT
AMAX1
RANF

Ways to Use This Material

This text material can be used in separate, distinct pieces—each piece corresponding to a few sections in some course text. Each chapter is written to be used as a separate unit. All the material presented has been used in classes, but never all in a single course. These text materials have been used in physical science major courses, engineering major courses, liberal arts physics courses, and premedical major courses. In all cases the students left the course with a better intuitive and calculational understanding of the material than was possible before the computer was used. Some of the material could not have been introduced without the computer. Project work on the computer has been especially effective. Sometimes different groups of students have developed different parts of a complicated computer routine; when the whole program is finally put together, everyone has gained from the shared experience.

Numerical Methods

Comments on numerical methods are deferred until the Appendix. After you have seen the use of the computer and have learned to use it to solve problems, you may want to investigate how to use numerical methods more systematically. Your sophistication with the computer will grow at your own rate.

Levels of the Material in this Book

Much of the physics in the undergraduate curriculum is covered. The accompanying chart illustrates which sections have been used at introductory, intermediate, and upper levels. The notation follows that of the Table of Contents, that is, 1.4 means Chapter 1, Section 4: two-dimensional motion.

	Introductory level		Intermediate	Upper undergraduate
Chapter	noncalculus	calculus	level	level
1	1.1	1.1	1.1	
	1.2	1.2	1.2	
	1.3	1.3	1.3	
	1.4	1.4	1.4	
	1.7	1.7	1.5	
			1.6	
			1.7	
			1.8	
2		2.1	2.1	
		2.2	2.2	

Chapter	Introductory level: noncalculus	Introductory level: calculus	Intermediate level	Upper undergraduate level
2		2.3	2.3	
3		3.1	3.1	3.1
		3.2	3.2	3.2
		3.3	3.3	3.3
				3.4
4		4.1	4.1	
		4.2	4.2	
		4.3	4.3	
		4.4	4.4	
		4.5	4.5	
5			5.1	5.1
			5.2	5.2
			5.3	5.3
			5.4	5.4
6		6.1	6.1	
		6.2	6.2	
		6.3	6.3	
		6.4	6.4	
		6.5	6.5	
7	7.1	7.1	7.1	
	7.2	7.2	7.2	
	7.3	7.3	7.3	
		7.4	7.4	
		7.5	7.5	
			7.6	
8	8.1	8.1	8.1	
	8.2	8.2	8.2	
	8.3	8.3	8.3	
		8.4	8.4	
		8.5	8.5	
		8.6	8.6	
		8.7	8.7	
9	9.1	9.1	9.4	
	9.2	9.2	9.5	
	9.3	9.3		
	9.4	9.4		
		9.5		
		9.6		
10		10.1	10.1	10.4
		10.2	10.2	10.5
		10.3	10.3	
			10.4	
			10.5	
11	11.1	11.1	11.1	
	11.2	11.2	11.2	

Chapter	Introductory level: noncalculus	Introductory level: calculus	Intermediate level	Upper undergraduate level
11	11.3	11.3	11.3	
	11.4	11.4	11.4	
		11.5	11.5	
		11.6	11.6	
			11.7	
12			12.1	12.1
			12.2	12.2
			12.3	12.3
			12.4	12.4
			12.5	12.5
				12.6
13	13.1	13.1	13.1	13.1
	13.2	13.2	13.2	13.2
	13.3	13.3	13.3	13.3
	13.4	13.4	13.4	13.4
14				14.1
				14.2
				14.3
				14.4
				14.5

Acknowledgments

Much of the material here was originally developed at Dartmouth College. Some of it is based originally on material developed for Project COEXIST. Project COEXIST was a National Science Foundation project at Dartmouth College to investigate the use of computers in engineering, mathematics, and physics. Two persons were directly involved with me in the physics section of Project COEXIST: Arthur W. Luehrmann and Elisha R. Huggins. A number of other people have also been helpful in the development of this book. Gustavus H. Zimmerman III, Norman L. Wentland, Robert C. Groman, Gregory P. Hughes, Peter B. Johnson, Albert B. Meador, John W. Merrill, Susan W. Merrill, and Kevin R. Squires have made especially valuable contributions. I would also like to thank three persons who were particularly helpful in the development of this book: Alfred Bork of the University of California, Irvine; David Gavenda of the University of Texas, Austin; and Robert J. Quigley of Western Washington State College.

John R. Merrill

1 mechanics

1.1 Introduction

This chapter introduces some of the uses of the computer in mechanics. For some of you, this is the first introduction to computing in physics education. The chapter begins at a simple level but ends up discussing some fairly sophisticated classical mechanics. Along the way you will learn some of the folklore about algorithmic (or iterative or step-by-step) calculations with the computer. The chapter starts with a discussion of one-dimensional motion and quickly turns to satellites. It continues with a discussion of the three-body problem and some classical scattering phenomena and ends with a discussion of coupled oscillators. All the examples given are applications of one basic $\mathbf{F} = m\mathbf{a}$ algorithm.

1.2 Basic Relationships

All the motions of classical systems are embodied in Newton's second Law, $\mathbf{F} = m\mathbf{a}$, and a few force laws, notably

$$\mathbf{F} = -G\frac{Mm}{r^3}\mathbf{r}$$

$$F = -mg$$

and

$$F = -ky$$

You will use the definitions of velocity ($\mathbf{v} \approx \Delta\mathbf{r}/\Delta\mathbf{t}$ for small $\Delta\mathbf{t}$) and acceleration ($\mathbf{a} \approx \Delta\mathbf{v}/\Delta\mathbf{t}$ for small $\Delta\mathbf{t}$).

A general feeling for Kepler's laws will help. An intuitive feeling for symmetry and how it affects solutions can be helpful, too.

All basic points are discussed in your physics textbook. You should have learned these basic facts from there.

1.3 One-Dimensional Motion

The basic problem of classical mechanics is simply stated: given a force law $\mathbf{F}(\mathbf{r}, \mathbf{v}, t)$ (a force $\mathbf{F}$ that may vary with position, velocity, and time) and initial conditions for positions and velocities, find the acceleration $\mathbf{a} = \mathbf{F}/m$ and, from the acceleration, determine the trajectory $\mathbf{r}(t)$ for the particle. This procedure can be solved analytically in some cases; you can then derive an equation for position as a function of time. Another way to solve the problem is by a straightforward step-by-step procedure. This second method is completely general, but it is easy to apply if you have a computer available. It involves a lot of arithmetic operations that would be tedious, although not impossible, to perform by hand. Let us state the step-by-step method in words.

1. Start at the initial point with the initial velocity.
2. Choose a time step Δt.
3. Calculate the force $\mathbf{F}$ and from it the acceleration $\mathbf{a}$.
4. Calculate the change in velocity $\Delta\mathbf{v} = \mathbf{a}\,\Delta\mathbf{t}$ and the change in position $\Delta\mathbf{r} = \mathbf{v}\,\Delta t$.
5. The new velocity and position are then $\mathbf{v} + \Delta\mathbf{v}$ and $\mathbf{r} + \Delta\mathbf{r}$, at the new time $t + \Delta t$.
6. Since you are now at the next point on the particle trajectory, return to step 3 and repeat 3, 4, and 5 (as often as you wish).

Using this procedure, you walk the particle along its trajectory a step at a time. The procedure would be exact if Δt were infinitesimal; since Δt is finite, the method can only be approximate. However, the approximations can be made very good by using a fairly small Δt and a modern computer carrying six- to ten-figure accuracy. (If Δt becomes too small, errors resulting from the fact that computers carry only finite accuracy become important.)

Programs in BASIC and FORTRAN which perform these calculations follow. The force on the particle is a uniform gravitational force in the $-y$ direction.

BASIC PROGRAM

```
 10 LET Y=100
 20 LET V=5
 30 LET M=2
 40 LET D=.1
100 LET F=-M*9.8
110 LET A=F/M
120 LET V=V+A*D
130 LET Y=Y+V*D
140 LET T=T+D
150 PRINT T,Y
160 IF T<5 THEN 100
999 END
```

FORTRAN PROGRAM

```
    Y=100.
    V=5.
    AM=2.
    DT=.1
 10 F=-AM*9.8
    A=F/AM
 16 V=V+A*DT
 17 Y=Y+V*DT
    T=T+DT
    PRINT 100,T,Y
 18 IF (T-5.) 10,20,20
 20 STOP
100 FORMAT (1X,2F 10.4)
    END
```

The first two lines set the initial conditions, the variables D, DT are the time step Δt. The step-by-step procedure based on $\mathbf{F} = m\mathbf{a}$ is bracketed. The force law (in this case just $F = -mg$) is written in line 100 (or statement 10 in the FORTRAN program); the acceleration is calculated as F/M. The new velocity V, position Y, and time T are calculated.

The BASIC statement in line 120 and the FORTRAN assignment statement for V should be read in the following way: take the old value of V, add to it $\Delta V (= a\ \Delta t)$ and store this new value as the present value of the variable V. (These instructions are assignment statements; the calculation on the right side of the equals sign is performed and the final value is assigned as the value of the variable on the left of the equals sign.) The PRINT instructions print the values as you go, and the program repeats the calculation until the particle has traveled for 5 s. The force law could be any (one-dimensional) force such as -K*Y for a spring, and so forth.

These programs can be executed on a computer. (The simplest method is called Euler's method.) You can get a better approximation to the true trajectory by using smaller and smaller Δt's (D or DT in the program). But smaller time steps mean more operations and hence more computer time. There is a simple way to get around this difficulty. In the jargon, it is called using a higher convergence method. The simplest such method is called the initial half step. You use a velocity v in the middle of the time step between t and $t + \Delta t$ when calculating the new position (Figure 1.1). (By keeping the accelerations in step with the positions, the accelerations are then automatically halfway between velocities.) [By the mean value theorem of calculus, there is some point in the (closed) interval $(t, t + \Delta t)$ such that the derivative at that point has the same slope as the chord between the old point y and the new point $y + \Delta y$ (for sufficiently smooth functions $y(t)$). You do not know how to find that point; but

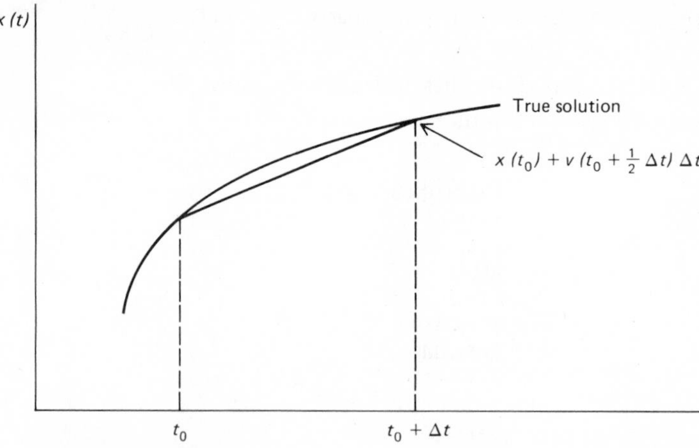

Figure 1.1 Comparison of true and numerical solutions using the initial half-step method.

usually a value of the derivative (in this case the velocity) near the center of the interval is a better approximation than the value of the derivative at the beginning of the interval.] By choosing the velocity at the midpoint of the interval (at $t + \frac{1}{2}\Delta t$), all you need to add to the programs to put the velocity at each step $\frac{1}{2}\Delta t$ ahead of the position (and acceleration) are the three lines

```
112 IF T>0 THEN 120                    IF (T) 15,15,16
114 LET V=V+A*D/2                   15 V=V+A*D/2
116 GOTO 130                           GOTO 17
```

These steps move v ahead by $\frac{1}{2}\Delta t$ the first time through the integration loop. Afterward, using a full Δt in each velocity calculation keeps the velocity $\frac{1}{2}$ step ahead.

Another useful addition to the simple program keeps the computer from printing out the results of every calculation it performs. As the program stands, when you change to a Δt one-tenth as big, you get 10 times as much print-out—an objectionable feature if you are sitting at a terminal. Add the three lines

```
142 LET DO=DO+D                        TPRIN = TPRIN + DT
144 IF DO<.2 THEN 160                  IF (TPRIN-.2) 18,19,19
146 LET DO=0                        19 TPRIN = 0.
```

These lines imply that the PRINT statements are reached only every 0.2 s—no matter what the time step. The final one-dimensional program might look like 1DFMA [Programs 1.1(a), (b)]. The acceleration $a = F/m$ is used, not the force.

Program 1.1(a)

1DFMABAS *Integrates 1D F = ma step-by-step*

```
10 LET Y=100        } Initialization
20 LET V=5          }
30 LET D=.02          Time step, Δt
40 LET A=-9.8         Acceleration
50 IF T>0 THEN 80   }
60 LET V=V+A*D/2    } Initial half-step
70 GOTO 90          }
80 LET V=V+A*D        New velocity
90 LET Y=Y+V*D        New position
100 LET T=T+D         New time
110 LET D0=D0+D     }
120 IF D0<.2 THEN 150 } Print
130 LET D0=0        } group
140 PRINT T,Y       }
150 IF T<5 THEN 40    Return for next Δt step
160 END
```

Program 1.1(b)

1DFMAFOR *Integrates 1D F = ma step-by-step*

```
      Y=100.          } Initialization
      VY=5.           }
      DELT=.02          Time step, Δt
      TPRIN=0.        } Initialization
      T=0.            }
5     AY=-9.8           Acceleration
      IF (T) 10,10,15 }
10    VY=VY+AY*DELT/2. } Initial half-step
      GOTO 20         }
15    VY=VY+AY*DELT     New velocity
20    Y=Y+VY*DELT       New position
      T=T+DELT          New time
      TPRIN=TPRIN+DELT                       }
      IF (TPRIN-(.2-DELT/2.)) 30,25,25       } Print
25    TPRIN=0.                               } group
      PRINT 100,T,Y                          }
100   FORMAT(1X,2F10.3)
30    IF (T-5.) 5,35,35   Return for next Δt step
35    STOP
      END
```

1.4 Two-Dimensional Motion

Two-dimensional motion is merely 2 one-dimensional motions at right angles. The program 2DFMA [Programs 1.2(a), (b), (c)] shows one way of calculating general two-dimensional step-by-step $\mathbf{F} = m\mathbf{a}$ integrations. (Satellite motion is the physical situation illustrated.) The force law is $\mathbf{F} = -m\mathbf{r}/|\mathbf{r}|^3$ (the universal law of gravitation, units scaled so that GM = 1). Only the acceleration matters to the motion so the force itself is not used. The results of the calculations are satellite orbits. Figure 1.2 shows a family of orbits calculated by means of the procedure shown in 2DFMA. All the orbits start at the same point $(-1, 0)$ but have different initial velocities $(0, v_y)$. The family members are ellipses with the circle having $v_y = 1$.

Program 1.2(a) Flowchart for a two-dimensional, iterative, **F** = m**a** strategy. After setting the initial values of parameters, the calculation walks along the trajectory of a particle a Δt step at a time. At each step, the components of the force on the particle determine the change in the velocity. The velocity determines the change in position.

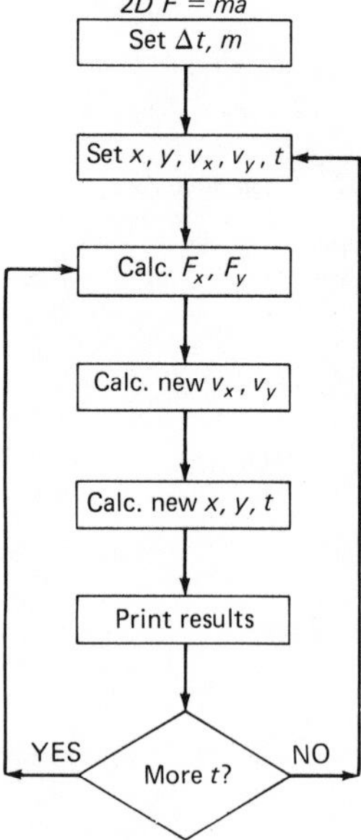

Program 1.2(b)

2DFMABAS *Integrates 2D F = ma step-by-step*

```
10 LET X=-1      }
20 LET Y=0       } Initialization
30 LET V1=0      }
40 LET V2=1      }
50 LET D=.01     Time step, Δt
60 LET R3=(X*X+Y*Y)↑1.5  }
70 LET A1=-X/R3   ax     } Acceleration
80 LET A2=-Y/R3   ay     }
90 IF T>0 THEN 130       }
100 LET V1=V1+A1*D/2     } Initial half-step
110 LET V2=V2+A2*D/2     }
120 GOTO 150             }
130 LET V1=V1+A1*D    New vx
140 LET V2=V2+A2*D    New vy
150 LET X=X+V1*D      New x
160 LET Y=Y+V2*D      New y
170 LET T=T+D    New time
180 LET T0=T0+D          }
190 IF T0<.5 THEN 220    } Print
200 LET T0=0             } group
210 PRINT T,X,Y          }
220 IF T<10 THEN 60   Return for next Δt step
230 END
```

One interesting thing to do with curves such as those in Figure 1.2 is to demonstrate Kepler's three laws. First, the orbits are ellipses. Find the second focus by symmetry [the force center is at (0, 0)]; calculate the sum of the distances from each point on

Program 1.2(c)

```
2DFMAFOR      Integrates 2D F = ma step-by-step

         X=-1.
         Y=0.
         VX=0.
         VY=1.                       Initialization
         T=0.
         PRNT=0.
         DELTAT=.005     Time step, 1/2 Δt
10       RADIUS=SQRT(X*X+Y*Y)
         RCUBED=RADIUS**3
         XACCEL=-X/RCUBED   a_x      Acceleration
         YACCEL=-Y/RCUBED   a_y
         VX=VX+XACCEL*DELTAT        New v_x
         VY=VY+YACCEL*DELTAT        New v_y
         DELTAT=.01      Reset time step to Δt
         X=X+VX*DELTAT        New x
         Y=Y+VY*DELTAT        New y
         T=T+DELTAT     New time
         PRNT=PRNT+DELTAT
         IF (PRNT-.5) 12,11,11            Print
11       PRNT=0.                          group
         PRINT 101,T,X,Y
12       IF (T-10.) 10,20,20     Return for next Δt step
101      FORMAT(1X,3E15.5)
20       STOP
         END
```

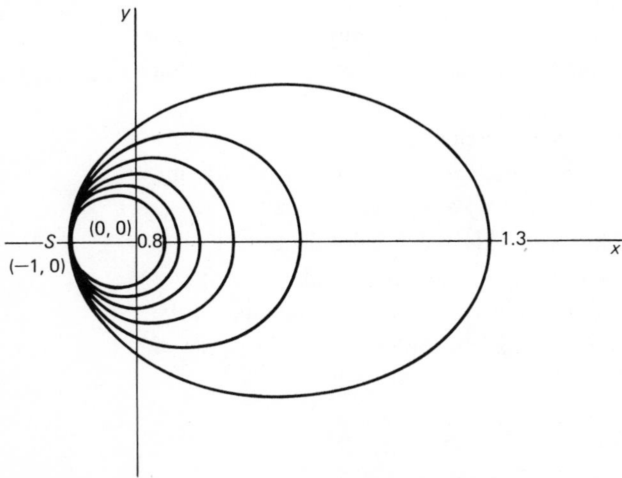

Figure 1.2 Family of orbits around a gravitational force center. All the orbits go through the point $S = (-1, 0)$ with velocities of 0.8, 0.9, 1, 1.1, 1.2, and 1.3 in the y direction. The units are normalized so that $\mathbf{a} = -\mathbf{r}/|\mathbf{r}|^3$.

the path to the two foci. If this sum is constant around the path, the trajectory is an ellipse with the force center at one focus. Second, the satellite sweeps equal areas in equal times. Since the time step is always the same, just print out the area swept in each step. The area (per unit time) is one-half the base of the triangle swept out times the height of the triangle, which is

$$\frac{1}{2}\,|\mathbf{r}\times\mathbf{v}| = \frac{1}{2}\sqrt{(yv_z - zv_y)^2 + (zv_x - xv_z)^2 + (xv_y - yv_x)^2}$$

Third, the period squared is proportional to the semimajor axis cubed. Measure the semimajor axis for several members of the family; find the period as twice the time it takes to go halfway around (the time to the first time y changes sign). Find the ratios $(\text{period})^2/(\text{semimajor axis})^3$ for several orbits and compare. (You may also be able to find that value analytically.)

1.5 Three-Body Motions

This strategy just discussed is not limited to two-body systems. You can do the same calculations for three bodies (or n bodies, as discussed in the chapter on statistical physics). Three-body motion is very complicated analytically; to test your program, you should limit yourself to symmetric three-body motions in a plane. The algorithmic method is completely general; you can start the three bodies in any way you wish (and the resulting motion is usually very complicated indeed).

The program 3BODY [Programs 1.3(a), (b)] shows one way to perform the calculations. The matrices R( ) and V( ) contain the positions and velocities of the three bodies; the masses are stored in the vector M( ). The initial values in the program 3BODY are chosen to keep everything very simple—the masses are equal and are started from the vertices of an equilateral triangle with equal velocities tangential to the circle through the positions. Again an initial half step is used. The force-law calculations calculate the force in pairs (both F_{ij}, the force on the ith body due to the jth body, and F_{ji} at once). The resulting positions are printed out only every one time unit.

Figure 1.3 shows a resulting set of orbits. When the magnitudes of the initial velocities are each exactly right, all three bodies move in the same circular orbit. (You can show this analytically by equating mv^2/r to the net force on each body due to the other two.) For other initial velocities (perpendicular to the lines from the center of the triangle), the orbits look like distorted beer signs.

1.6 Classical Scattering

Another interesting application of the $\mathbf{F} = m\mathbf{a}$ strategy is to (classical) scattering phenoma. The comet problem is much like the satellite problem discussed above; but you

Program 1.3(a)

```
3BODYBAS     Integrates F = ma for three bodies in 3D

100 DIM R(3,3),V(3,3),M(3)
110 READ M(1),M(2),M(3)   Masses
120 DATA 1,1,1
130 LET D0=.02       Time step, Δt
140 LET D=D0/2       Time step (initially Δt/2)
150 READ R(1,1),R(2,1),R(3,1)  }
160 DATA 1,-1,0                }
170 READ R(1,2),R(2,2),R(3,2)  }  Initial positions
180 DATA 0,0,1.73205           }  (equilateral triangle)
190 READ R(1,3),R(2,3),R(3,3)  }
200 DATA 0,0,0                 }
210 READ V(1,1),V(2,1),V(3,1)           }
220 DATA .353553,.353553,-.707107       }
230 READ V(1,2),V(2,2),V(3,2)           }  Initial
240 DATA .612372,-.612372,0             }  velocities
250 READ V(1,3),V(2,3),V(3,3)           }
260 DATA 0,0,0                          }
270 FOR I=1 TO 3                                      }
280 FOR J=I+1 TO 3                                    }
290 LET X=R(J,1)-R(I,1)      }                        }
300 LET Y=R(J,2)-R(I,2)      }  r_ij                  }
310 LET Z=R(J,3)-R(I,3)      }                        }
320 LET R=SQR(X*X+Y*Y+Z*Z)   }                        }
330 LET F1=-M(I)*M(J)*X/(R*R*R)  }                    }
340 LET F2=-M(I)*M(J)*Y/(R*R*R)  }  F_ij              }  New
350 LET F3=-M(I)*M(J)*Z/(R*R*R)  }                    }  velocities
360 LET V(J,1)=V(J,1)+F1*D/M(J)   New v_jx            }
370 LET V(I,2)=V(J,2)-F1*D/M(I)   New v_ix            }
380 LET V(J,2)=V(J,2)+F2*D/M(J)   New v_jy            }
390 LET V(I,2)=V(I,2)-F2*D/M(I)   New v_iy            }
400 LET V(J,3)=V(J,3)+F3*D/M(J)   New v_ji            }
410 LET V(I,3)=V(I,3)-F3*D/M(I)   New v_ix            }
420 NEXT J                                            }
430 NEXT I                                            }
440 LET D=D0     Reset time step to Δt
450 FOR I=1 TO 3                      }
460 FOR J=1 TO 3                      }
470 LET R(I,J)=R(I,J)+V(I,J)*D        }  New
480 NEXT J                            }  positions
490 NEXT I                            }
500 LET T=T+D    New time
510 LET T0=T0+D          }
520 IF T0<1 THEN 620     }
530 LET T0=0             }
540 PRINT "T =";T        }  Print
550 LET T0=0             }  group
560 FOR I=1 TO 3         }
570 FOR J=1 TO 3         }
580 PRINT R(I,J),        }
590 NEXT J               }
600 PRINT                }
610 NEXT I               }
620 IF T<10 THEN 270     Return for next Δt step
630 END
```

Program 1.3(b)

3BODYFOR *Integrates F = ma for three bodies in 3D*

```
      DIMENSION RAD(3,3),VEL(3,3),FMASS(3)
      DATA FMASS /1.,1.,1./   Masses
      TIMINT=.02   Time step, Δt
      TIM1=TIMINT/2.   Time step, Δt/2
      DATA RAD(1,1),RAD(2,1),RAD(3,1)/1.,-1.,0./              }
      DATA RAD(1,2),RAD(2,2),RAD(3,2)/0.,0.,1.73205/          } Initial positions
      DATA RAD(1,3),RAD(2,3),RAD(3,3)/0.,0.,0./               }
      DATA VEL(1,1),VEL(2,1),VEL(3,1)/.353553,.353553,-.707107/ } Initial
      DATA VEL(1,2),VEL(2,2),VEL(3,2)/.612372,-.612372,0./      } velocities
      DATA VEL(1,3),VEL(2,3),VEL(3,3)/0.,0.,0./                 }
5     DO 20 I=1,2                                                 }
      DO 10 J=(I+1),3                                             }
      X=RAD(J,1)-RAD(I,1)    }                                    }
      Y=RAD(J,2)-RAD(I,2)    } r_ij                               }
      Z=RAD(J,3)-RAD(I,3)    }                                    }
      R=SQRT(X**X+Y**Y+Z**Z) }                                    }
      FX=-FMASS(I)**FMASS(J)**X/(R**R**R) }                       } New
      FY=-FMASS(I)**FMASS(J)**Y/(R**R**R) } F_ij                  } velocities
      FZ=-FMASS(I)**FMASS(J)**Z/(R**R**R) }                       }
      VEL(J,1)=VEL(J,1)+FX**TIM1/FMASS(J)   New v_jx              }
      VEL(I,1)=VEL(I,1)-FX**TIM1/FMASS(I)   New v_ix              }
      VEL(J,2)=VEL(J,2)+FY**TIM1/FMASS(J)   New v_jy              }
      VEL(I,2)=VEL(I,2)-FY**TIM1/FMASS(I)   New v_iy              }
      VEL(J,3)=VEL(J,3)+FZ**TIM1/FMASS(J)   New v_jz              }
10    VEL(I,3)=VEL(I,3)-FZ**TIM1/FMASS(I)   New v_iz              }
20    CONTINUE
      TIM1=TIMINT   Reset time step to Δt  }
      DO 40 I=1,3                          } New
      DO 30 J=1,3                          } positions
30    RAD(I,J)=RAD(I,J)+VEL(I,J)**TIM1     }
40    CONTINUE
      TIM=TIM+TIMINT               New time
      TIMO=TIMO+TIMINT                   }
      IF(TIMO-1.)60,50,50                }
50    TIMO=0.                            } Print group
      PRINT 100,TIM                      }
      DO 80 I=1,3                        }
80    PRINT 110,(RAD(I,J),J=1,3)         }
60    IF (TIM-10.) 5,90,90   Return for next Δt step
90    STOP
100   FORMAT(1X,3HT =,F10.6)
110   FORMAT(1X,3F10.6)
      END
```

must be careful that the step-by-step procedure works even when the comet comes very close to the force center. Figure 1.4 shows a series of comet trajectories.

A particularly interesting example of classical scattering is the scattering of point electrons and positrons off a simple model of a hydrogen atom. (The true physical scattering in such a situation is quantum mechanical because the atom is small.)

Figure 1.5 shows trajectories of positrons off the model of hydrogen. The hydrogen atom is treated as a positive point nucleus surrounded by a uniform, negatively charged cloud with uniform density (the electron). The total negative charge in the cloud equals the positive charge on the point nucleus. Outside the cloud (the dashed circle

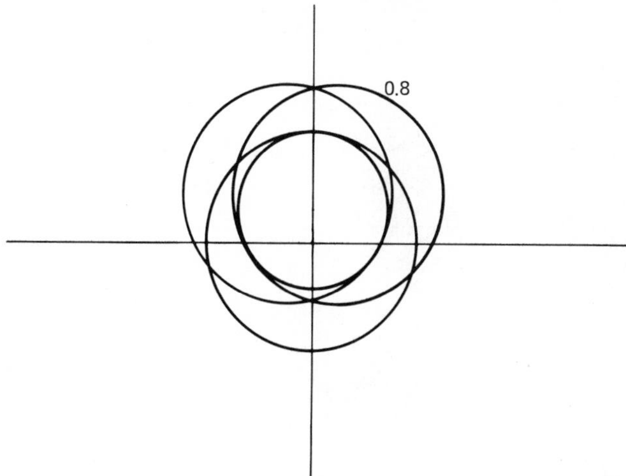

Figure 1.3 Orbits for three-body motion. The masses are equal; the speeds are equal; and the initial positions lie on the corners of one equilateral triangle. The velocities are all perpendicular to the line from the center.

in the figure), the force on the incoming positron is zero. At any point inside the cloud, part of the repulsion between incoming positron and the positive nucleus is screened by the negative cloud. Several trajectories of positrons are shown. From figures like this (but with more trajectories drawn in), you can count the number of positrons deflected by 0° to 10°, 10° to 20°, and so forth.

Figure 1.6 shows some trajectories of electrons scattering off a model of a hydrogen atom. Now the incoming particles are attracted to the nucleus, and strange looping trajectories occur. The electrons must always ultimately leave the system because they have positive total energy. Under some circumstances, it may take a long time for the electron to get away. Figure 1.7 shows a few looping trajectories near the critical value of angular momentum and energy at which looping occurs. [For more detail on these questions, see J. R. Merrill and R. A. Morrow, *Am. J. Phys.*, **38**, 1104 (1970).]

The programming to perform these scattering calculations is somewhat more complicated than that in 2DFMA. In particular, a very high convergence method must be used so that the trajectory will be correct even very near the force center. It sometimes turns out to be useful to use a variable time step so that (approximately) the impulse $\mathbf{F}\,\Delta t$ at each step is about the same. The program SCAT [Programs 1.4(a), (b)] is one possible way to do the calculations.

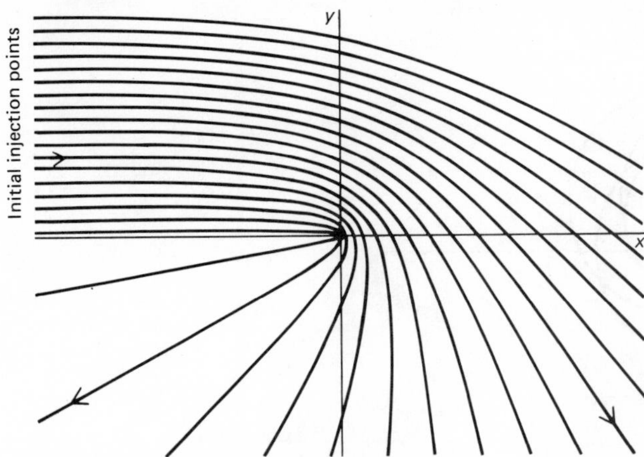

Figure 1.4 Several trajectories for unbound orbits around a point, gravitational force center. The trajectories all start with the same (horizontal) velocity but have different impact parameters (initial vertical positions).

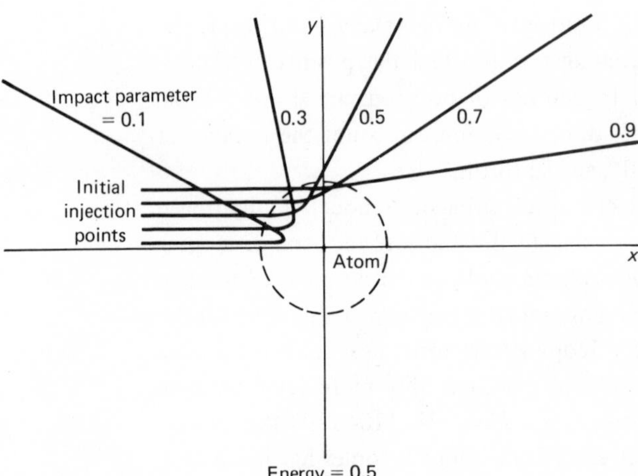

Figure 1.5 Trajectories of (classical) positrons off a model of atomic hydrogen. The atom is a point nucleus surrounded by a uniform negative cloud out to the cloud circle. The initial kinetic energies equal 0.5 for the five angular momenta shown.

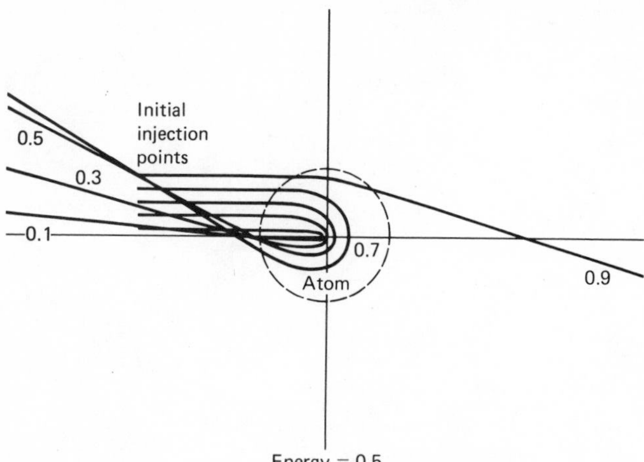

Figure 1.6 Trajectories of (classical) electrons off a model of a hydrogen atom. The looping trajectories occur because the force is now attractive. The curves are marked by their initial y positions (the impact parameters).

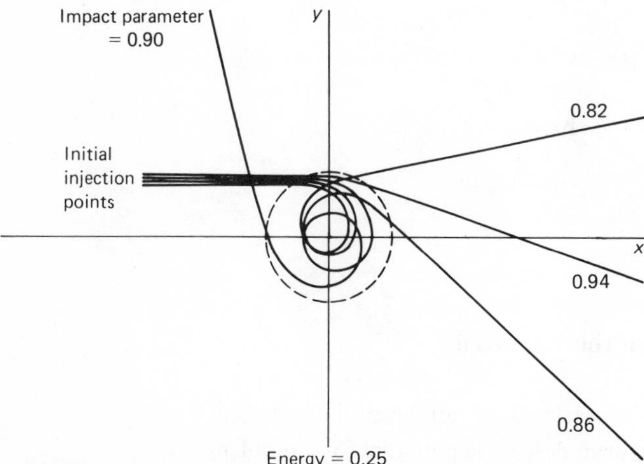

Figure 1.7 Several closely spaced looping orbits for electrons off hydrogen model. The initial kinetic energies are all 0.25; the curves are marked by their initial y positions.

Program 1.4(a)

SCATBAS *Scatters particles off force center*

```
100 PRINT "INITIAL (X,Y) & ENERGY";
110 INPUT X,Y,E
120 LET Q=-1  q/m
130 LET V1=SQR(2*E)   } Initial
140 LET V2=0          } velocity
150 LET B1=0   }
160 LET B2=0   }
170 LET B3=0   }
180 LET C1=0   } Initialization
190 LET C2=0   }
200 LET C3=0   }
210 LET D=.05  }
220 LET R=SQR(X*X+Y*Y)
230 IF R>1 THEN 430   If far enough away, leave alone
240 LET F1=Q*(X/(R*R*R)-X)   First a_x
250 LET G1=Q*(Y/(R*R*R)-Y)   First a_y
260 LET D=.05/(1+SQR(F1*F1+G1*G1))   Time step, Δt (variable)
270 LET X1=X+V1*D/2
280 LET Y1=Y+V2*D/2
290 LET R=SQR(X1*X1+Y1*Y1)          Method fits to polynomial
300 LET F2=Q*(X1/(R*R*R)-X1)        a_x to b_1 + b_2 Δt + b_3(Δt)^2
310 LET G2=Q*(Y1/(R*R*R)-Y1)        a_y to c_1 + c_2 Δt + c_3(Δt)^2
320 LET X1=X+V1*D                   in each interval
330 LET Y1=Y+V2*D                   by using values
340 LET R=SQR(X1*X1+Y1*Y1)          at t, t + 1/2 Δt, t + Δt;
350 LET F3=Q*(X1/(R*R*R)-X1)        then integrates polynomial
360 LET G3=Q*(Y1/(R*R*R)-Y1)
370 LET B1=F1
380 LET B3=4*(B1/2-F2+F3/2)/(D*D)
390 LET B2=2*(F2-B1-B3*D*D/4)/D
400 LET C1=G1
410 LET C3=4*(C1/2-G2+G3/2)/(D*D)
420 LET C2=2*(G2-C1-C3*D*D/4)/D
430 LET X=X+V1*D+B1*D*D/2+B2*D*D*D/6+B3*D*D*D*D/12   New x
440 LET Y=Y+V2*D+C1*D*D/2+C2*D*D*D/6+C3*D*D*D*D/12   New y
450 LET V1=V1+B2*D*D/2+B3*D*D*D/3+B1*D   New v_x
460 LET V2=V2+C2*D*D/2+C3*D*D*D/3+C1*D   New v_y
470 LET T=T+D   New time
480 LET T0=T0+D       }
490 IF T0<1 THEN 520  } Print
500 LET T0=0          } group
510 PRINT T;X;Y       }
520 IF T<10 THEN 150   Return for next Δt step
530 PRINT
540 GOTO 100   Return for new trajectory
550 END
```

1.7 Two-Dimensional Motion Given the Potential

You can also solve for two- (or more-) dimensional motion given the potential $V(\mathbf{r})$ rather than the force $\mathbf{F}(\mathbf{r})$. Often it is easier to derive the potential for a problem than to derive the force itself (particularly in problems in electrostatics). The relationship between the potential V and the force is just $F = -\text{grad}(V)$, where the grad is a vector derivative. [In Cartesian coordinates $\text{grad}(V(x, y, z)) = (\partial V/\partial x, \partial V/\partial y, \partial V/\partial z)$; partial derivatives are just like regular derivatives but with the other variables held constant.]

Program 1.4(b)

```
SCATFOR    Scatters particles off force center

5      READ(5,110)X,Y,E    Initial position, energy
       QTOM=-1.   q/m
       VX=SQRT(2.*E)    } Initial
       VY=0.            } velocity
       TIM=0.   }
       TPR=0.   }
10     AX1=0.   }
       AX2=0.   }
       AX3=0.   } Initialization
       AY1=0.   }
       AY2=0.   }
       AY3=0.   }
       DT=.05   Time step, Δt
       R=SQRT(X*X+Y*Y)
       IF (R-1.) 20,20,30    If far enough away, leave alone
20     F1X=QTOM*(X/(R*R*R)-X)    First a_x
       F1Y=QTOM*(Y/(R*R*R)-Y)    First a_y
       DT=.05/(1.+SQRT(F1X*F1X+F1Y*F1Y))    Time step, Δt (variable)
       X1=X+VX*DT/2.
       Y1=Y1+VY*DT/2.                       Method fits to polynomial
       F2X=QTOM*(X1/(R*R*R)-X1)             a_x: ax_1 + ax_2(Δt) + ax_3(Δt)^2
       F2Y=QTOM*(Y1/(R*R*R)-Y1)             a_y: ay_1 + ay_2(Δt) + ay_3(Δt)^2
       X1=X+VX*DT                           in each interval
       Y1=Y+VY*DT                           by using values
       R=SRQT(X1*X1+Y1*Y1)                  at t, Δt + 1/2 Δt, t + Δt;
       F3X=QTOM*(X1/(R*R*R)-X1)
       F3Y=QTOM*(Y1/(R*R*R)-Y1)             then integrates polynomial
       AX1=F1X
       AX3=4.*(AX1/2.-F2X+F3X/2.)/(DT*DT)
       AX2=2.*(F2X-AX1-AX3*DT*DT/4.)/DT
       AY1=F1Y
       AY3=4.*(AY1/2.-F2Y+F3Y/2.)/(DT*DT)
       AY2=2.*(F2Y-AY1-AY3*DT*DT/4.)/DT
       X=X+VX*DT+AX1*DT*DT/2.+AX2*DT*DT*DT/6.+AX3*DT*DT*DT*DT/12.  New x
       Y=Y+VY*DT+AY1*DT*DT/2.+AY2*DT*DT*DT/6.+AY3*DT*DT*DT*DT/12.  New y
       VX=VX+AX1*DT+AX2*DT*DT/2.+AX3*DT*DT*DT/3.    New v_x
       VY=VY+AY1*DT+AY2*DT*DT/2.+AY3*DT*DT*DT/3.    New v_y
       TIM=TIM+DT    New time
       TPR=TPR+DT                }
       IF (TPR-1.) 40,30,30      } Print
30     TPR=0.                    } group
       PRINT 120,TIM,X,Y         }
40     IF (TIM-10.) 10,5,5    Return for next Δt step or new trajectory
110    FORMAT(3F10.6)
120    FORMAT(1X,3F10.6)
       END
```

When the problem is spherically symmetric, that is, when the potential (and the force) depend only on $r = |\mathbf{r}|$ (central-force problems), the relationship between **F** and V is even simpler:

$$\mathbf{F}(\mathbf{r}) = -\frac{dV(r)}{dr}\frac{\mathbf{r}}{|r|}$$

When the potential is spherically symmetric, you simply take the derivative of the potential with respect to r and multiply by a unit radial vector. In a program you can find the derivative dV/dr numerically by evaluating the potential V at two points (r and $r + \Delta r$) close together and then dividing by Δr.

Some simple examples of potentials which occur often in physics are: $V(r) = -a/r$, the Coulomb potential for the inverse-square force law; $V(r) = -a \exp(-r/r_0)/r$, the screened-Coulomb potential; and $V(r) = a/r^{12} - b/r^6$, the 6–12 or Lennard-Jones potential. a, b, and r_0 are constants which depend on the particular problem. The Coulomb potential is appropriate for gravitational and point-charge electrostatic problems. The screened-Coulomb potential arises in many-electron problems. The Lennard-Jones potential arises in molecular and in solid-state physics problems.

The program 2DV [Programs 1.5(a), (b)] computes two-dimensional trajectories given the potential $V(r)$.

1.8 Coupled Oscillators

The last application of the **F** = m**a** computer strategy to mechanics has to do with coupled oscillators. Often two physical systems which oscillate are sufficiently coupled so that motion in one produces motion in the other. One picture is of two masses hung on springs with a third spring between them (Figure 1.8).

Program 1.5(a)

```
2DVBAS    Integrates 2D F = ma given the potential, V(r)

100 DEF FNV(R)=-1/R     Potential V(r)
110 DEF FNF(R)=-(FNV(R+1E-4)-FNV(R-1E-4))/2E-4     Force/mass
120 PRINT "INITIAL X,Y,VX,VY";  }
130 INPUT X,Y,V1,V2             }
140 LET T=0                     }
150 LET T1=0                    } Initialization
160 LET D=1/32   Time step, Δt  }
170 LET T0=5     Final time     }
180 PRINT T,X,Y,V1,V2
190 LET R=SQR(X*X+Y*Y)  }
200 LET A=FNF(R)        } Acceleration
210 LET A1=A*X/R   ax   }
220 LET A2=A*Y/R   ay   }
230 IF T>0 THEN 270     }
240 LET V1=V1+A1*D/2    } Initial half-step
250 LET V2=V2+A2*D/2    }
260 GOTO 290            }
270 LET V1=V1+A1*D   New vx
280 LET V2=V2+A2*D   New vy
290 LET X=X+V1*D   New x
300 LET Y=Y+V2*D   New y
310 LET T=T+D   New time
320 LET T1=T1+D             }
330 IF T1<T0/16 THEN 360    } Print
340 LET T1=0                } group
350 PRINT T,X,Y,V1,V2       }
360 IF T<T0 THEN 190   Return for next Δt step
370 PRINT
380 GOTO 120   Return for new trajectory
390 END
```

Program 1.5(b)

```
2DVFOR    Integrates 2D F = ma given the potential V(r)

          POTEN(R)=-1/R    Potential V(r)
          F(R)=-(POTEN(R+.0001)-POTEN(R-.0001))/.0002   Force/mass
5         READ(5,100)X,Y,VX,VY  ┐
          IF (X-999.) 7,45,45   │
7         T=0.                  ├Initialization
          PRNTT=0.              │
          DELT=1./32.   Δt      │
          ENDT=6.    Final time ┘
          PRINT 101,T,X,Y,VX,VY
──10      R=SQRT(X*X+Y*Y) ┐
│         A=F(R)          ├Acceleration
│         AX=A*X/R   a_x  │
│         AY=A*Y/R   a_y  ┘
│         IF (T) 15,15,20 ┐
│ 15      VX=VX+AX*DELT/2.├Initial half-step
│         VY=VY+AY*DELT/2.│
│         GOTO 25         ┘
│ 20      VX=VX+AX*DELT   New v_x
↑         VY=VY+AY*DELT   New v_y
│ 25      X=X+VX*DELT     New x
│         Y=Y+VY*DELT     New y
│         T=T+DELT        New time
│         PRNTT=PRNTT+DELT                 ┐
│         IF (PRNTT-ENDT/16.) 40,30,30     ├Print
│ 30      PRNTT=0.                         │group
│         PRINT 101,T,X,Y,VX,VY            ┘
└─40      IF (T-ENDT) 10,5,5   Return for next Δt step or new trajectory
45        STOP
101       FORMAT(1X,5F10.4)
          END
```

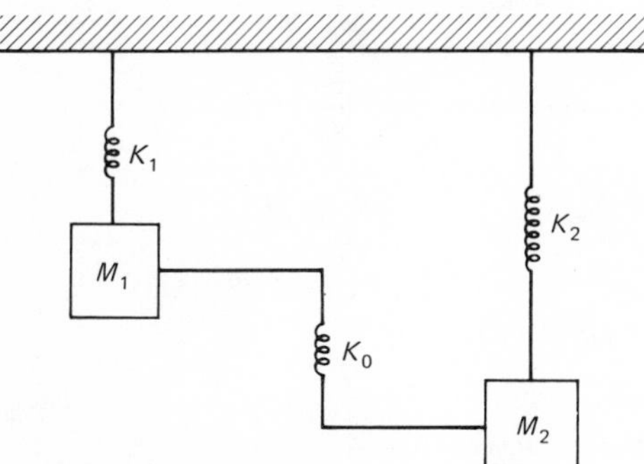

Figure 1.8 Mass and spring system for two, coupled, vertical oscillators.

The force laws are

$$F_1 = -k_1 y_1 - k_0(y_1 - y_2) \qquad F_2 = -k_2 y_2 - k_0(y_2 - y_1)$$

y_1 and y_2 are the displacements of m_1 and m_2 from equilibrium. The equations can be solved directly by the $\mathbf{F} = ma$ numerical technique. The program COUPLE [Programs 1.6(a), (b)] is one way to do it. The simple case of $m_1 = m_2 = k_1 = k_2 = 1$ (with $k_0 = 0.1$) is shown, but the program is general.

Normal modes are easy to illustrate with the program. A normal mode is a motion for the system in which energy (or motion) is not traded back and forth between the masses. For two one-dimensional oscillators, there are two normal modes; by symmetry you can guess that (for the special case of $m_1 = m_2 = k_1 = k_2 = 1$) the two normal modes can start with $y_1 = 1$, $y_2 = 1$, $v_1 = v_2 = 0$, or $y_1 = -1$, $y_2 = 1$, $v_1 = v_2 = 0$. You can show with the program that these modes do not trade motion or energy between the masses. Any general motion (started from less-symmetric initial conditions) can be written analytically as a linear combination of the normal mode motions.

The method works for any number of oscillators of any complexity. The coupling need not be by a simple spring. In fact, the oscillators need not be simple harmonic oscillators, so the method is an introduction to the general analysis of complicated coupled systems. The idea of normal modes is not useful except when the forces are (nearly) harmonic.

This chapter was a brief introduction to a few of the ways a simple, algorithmic

Program 1.6(a)

COUPLEBA *Motion of coupled oscillators*

```
100 READ K1,M1,K2,M2,K0            Spring constants
110 DATA 1,1,1,1,.1                and masses
120 PRINT "INITIAL (Y1,V1,Y2,V2)";           }
130 INPUT Y1,V1,Y2,V2                        }
140 LET T=0                                  }
150 LET T0=0                                 } Initialization
160 LET D=.005    Time step, Δt/2            }
170 LET D0=.01    Time step stored           }
180 LET A1=(-K1*Y1-K0*(Y1-Y2))/M1   Acceleration of first mass
190 LET A2=(-K2*Y2-K0*(Y2-Y1))/M2   Acceleration of second mass
200 LET V1=V1+A1*D      New v of first mass
210 LET V2=V2+A2*D      New v of second mass
220 LET D=D0    Full time step
230 LET Y1=Y1+V1*D      New position of first mass
240 LET Y2=Y2+V2*D      New position of second mass
250 LET T=T+D       New time
260 LET T0=T0+D                   }
270 IF T0<3.14159/4 THEN 300      } Print
280 LET T0=0                      } group
290 PRINT T;Y1;Y2                 }
300 IF T<4*3.14159 THEN 180    Return for next Δt step
310 PRINT
320 GOTO 120      Return for new initial conditions
330 END
```

Program 1.6(b)

```
COUPLEFO     Motion of coupled oscillators

        DATA AK1,AM1,AK2,AM2,AK0/1.,1.,1.,1.,.1/  Spring constants and masses
5       READ(5,100)Y1,VY1,Y2,VY2     Initial conditions
        IF (Y1-999.) 7,50,7
7       T=0.                                              Initialization
        PRNTT=0.
        DELT=.01     Time step, Δt
        DELTT=DELT/2.     Initial time step, 1/2 Δt
10      AY1=(-AK1*Y1-AK0*(Y1-Y2))/AM1     Acceleration of first mass
        AY2=(-AK2*Y2-AK0*(Y2-Y1))/AM2     Acceleration of second mass
        VY1=VY1+AY1*DELTT     New velocity of first mass
        VY2=VY2+AY2*DELTT     New velocity of second mass
        DELTT=DELT     Full time step
        Y1=Y1+VY1*DELTT     New position of first mass
        Y2=Y2+VY2*DELTT     New position of second mass
        T=T+DELTT     New time
        PRNTT=PRNTT+DELTT
        IF (PRNTT-3.14159/4.) 20,15,15     Print
15      PRNTT=0.                           group
        PRINT 101,T,Y1,Y2
20      IF (T-4.*3.14159) 10,5,5     Return for next Δt step or new initial conditions
50      STOP
100     FORMAT(4F10.4)
101     FORMAT(1X,3F10.4)
        END
```

F = ma computer strategy can be used to advantage in mechanics. Even the more complicated particle mechanics problems can be treated in a straightforward way. The advantages of the method are its generality and the ease with which it can be understood. The major disadvantage is that global properties of the solutions (such as conservation laws) are not immediately evident in the solutions. Conservation laws can easily be illustrated from the solutions and often form sensitive tests of the accuracy of the numerical method, but such laws are not derived by numerical calculations.

The general algorithmic approach, that of having the computer walk along a curve, is a strategy you will see again in later chapters. The method is not unique to the computer; the calculations can be done by hand. However, the computer can add, multiply, and so forth, much faster (and more accurately) than an individual can, and the computer does not get bored doing the same operations over and over.

Problems

The following are illustrative of the kinds of exercises that have been useful in relation to the material discussed in the chapter.

1. Using a program implementing the two-dimensional **F** = ma strategy for a single gravitational force center at the origin (the satellite problem),
   a. Find five members of the family of closed orbits going through $(-2, 0)$ with various velocities $(0, v_y)$.

b. For one of these orbits show that the orbit is an ellipse with the force center at one focus.
c. For one of these orbits show that equal areas are swept out in equal times Δt.
d. Show that the square of the period is proportional to the cube of the semimajor axis, using several of the orbits.

2. Using a program implementing the two-dimensional $\mathbf{F} = m\mathbf{a}$ strategy for a single gravitational force center at the origin,
   a. Find five members of the family of orbits going through $(-1,0)$ having the same total energy at $(-1,0)$ but different angular momenta. (These are orbits having the same magnitude of velocity but different directions of velocity at the initial point.)
   b. Demonstrate Kepler's three laws for these orbits.

3. Using a program implementing the two-dimensional $\mathbf{F} = m\mathbf{a}$ strategy for two, fixed, equal-mass, gravitational force centers at $(0, 0)$ and $(10, 0)$,
   a. Find an orbit that loops outside both force centers and returns (more or less), starting at the point $(-1, 0)$.
   b. Find a figure-eight orbit which loops around both force centers, starting at the point $(-1, 0)$.
   c. Calculate the difference in total energy between the two orbits from parts a and b.

4. In a system of units which keeps the gravitational force law simple,
   a. Find the distance between the earth and the moon, the masses of the earth and the moon, and the distance from the center of the earth to a circular orbit 100 mi above the surface of the earth.
   b. Find the correct initial velocities (wholly in the y direction) that provide orbits which loop the moon and return to earth, starting with a satellite 100 mi above the surface of the earth. Find one orbit looping outside both the earth and the moon and one figure-eight orbit.
   c. Find out what the energy difference between the two orbits from part b is. Why did the Apollo program (particularly Apollo 8) use the figure-eight "save" orbit?

5. Given three equal-mass bodies interacting by the gravitational force and lying initially on the corners of an isosceles triangle,
   a. Find three sets of initial velocities which give interesting and differing motions for the three bodies.
   b. Examine the conservation of total energy and total angular momentum of the system, for each set of initial velocities.
   c. Examine the kinetic, potential, and total energies and the angular momentum as time progresses for each set of initial velocities and for one body.

6. Given the Yukawa potential $V(r) = -\exp(-r)/r$ (where the force center is at the origin),
   a. Find several orbits going through the point (2, 0) using a two-dimensional $\mathbf{F} = m\mathbf{a}$ program (based on potentials).
   b. Find the correct velocity to produce a circular orbit through (2, 0) using the program.
   c. Determine what your velocity from part b should have been by equating the force due to the Yukawa potential and the centripetal force. Compare the two results.

7. Given the Yukawa potential $V(r) = -\exp(-r)/r$,
   a. Start particles moving parallel to the x axis at several positions $(-5, y)$, using a two-dimensional $\mathbf{F} = m\mathbf{a}$ program based on the potential. Compute the trajectories and measure the angles the final trajectories (leaving the system) make with the x direction (the initial trajectory direction.)
   b. Plot the number of particles scattered by 0° to 10°, 10° to 20°, etc., versus angle of deviation.
   c. Repeat the trajectory computation for $V(r) = -\exp(-r/5)/r$ and show that the plot from b changes.
   d. Explain how trajectories of scattered particles can give information on the details of the potential.

8. Consider the coupled spring system shown in Figure 1.9.
   a. Write down the (coupled) equations of motion for the x motions of the two (equal) masses.
   b. Using an $\mathbf{F} = m\mathbf{a}$ program find the motions that result when you start the masses in the following ways:
      (1) Both masses started from rest having been pulled equal amounts to the right.
      (2) Both masses started from rest having had the masses pulled equal amounts in opposite directions.
      (3) One mass started at its equilibrium point and the other displaced to the right.
   c. Now allow the springs to be nonlinear so that $F = -k(x + 0.1x^3)$. Repeat part b and compare.

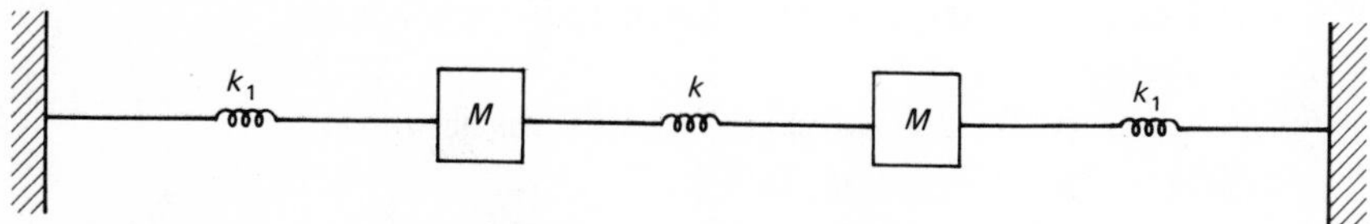

Figure 1.9 Mass and spring system for two, coupled, horizontal oscillators.

2 thermodynamics

2.1 Introduction

This short chapter shows ways of using the computer to advantage in thermodynamics. The computer makes it easy to illustrate reversible processes performed on many materials. The simplest material is an ideal gas, which is the illustration used in this chapter. The way the computer is used in this chapter is entirely general, however; it can be applied equally well to real substances and complicated *PVT* equations of state. The *PVT* relation can even be a table of numbers rather than an analytic expression.

2.2 *PVT* Surfaces

Every material has an equilibrium relationship between pressure P, volume V, and temperature T. In general, for a real substance the relationship is quite complicated, since in some ranges of P, V and T, the substance will be a solid and in others, a liquid or a gas.

A useful idealization for introductory use is the ideal gas. An ideal gas has the simple equation of state

$$PV = nRT$$

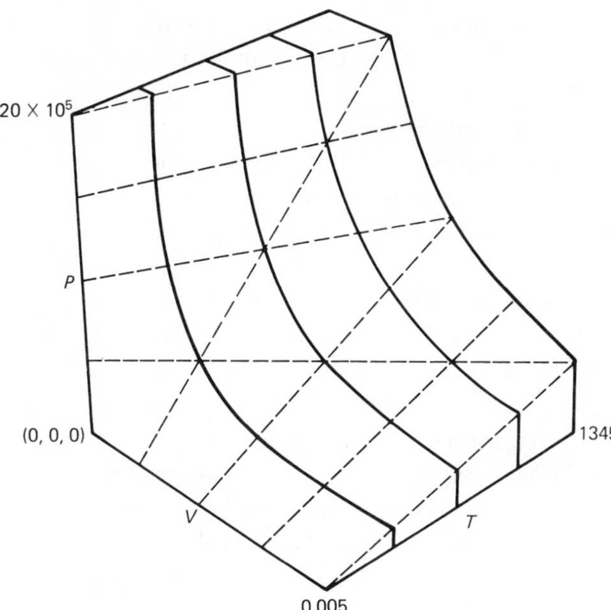

Figure 2.1 *PVT* surface of an ideal gas. The contours shown are isotherms (constant temperature), isobars (constant pressure), and isochors (constant volume). All units are in the mks system.

where R is the universal gas constant (8.314×10^3 J/kg · mole · deg) and n is the number of moles of gas present. P is the pressure in newtons per square meter; V is the volume in cubic meters; and T is the absolute temperature in degrees Kelvin. The ideal gas is useful because its *PVT* relation is so simple. On the other hand, all sufficiently dilute gases are good approximations to an ideal gas. Since the ideal gas does not condense into a liquid or a solid, it cannot represent any real substance over large regions of P, V, or T.

Figure 2.1 shows part of the *PVT* relationship for the ideal gas. *mks* units are used. Pressure is one axis, with one atmosphere equal to 1.013×10^5 N/m^2. Volume is a second axis, with volumes up to 0.005 m^3 shown. Temperature is the third axis, with T's up to about 1350 K. Constant-pressure contours (isobars) are straight lines, $V \propto T$. Constant-volume contours (isochors) are also straight lines, $P \propto T$. Constant-temperature contours (isotherms) are rectangular hyperbolas, $P \propto 1/V$.

The *PVT* surface represents all the possible equilibrium states or situations that the material (in this case an ideal gas) can have. Reversible processes, which the material undergoes, are a succession of equilibrium states. So all reversible processes are lines of some sort on the *PVT* surface. Irreversible processes need not be a succession of equilibrium states and so may not be represented as any contour on the equilibrium *PVT* surface.

Many properties of the substance, notably the internal energy U and the entropy S, are functions of the state of the substance, which means that U and S have unique values (up to an additive constant just like potential energy) at each point in the *PVT* surface. The changes in U and S, when the system undergoes some process, depend only on the initial and final states.

The heat Q exchanged between the material and its environment depends on the initial and final states and also on precisely how the system goes from one state to the other. The work W done by or on the system is also path-dependent. One of the major tasks of thermodynamics is to use the empirical laws of thermodynamics to find Q, W, ΔU, and ΔS for processes performed on real substances. This chapter concentrates on reversible processes performed on an ideal gas because these processes are easiest to understand. However, the computer can be applied perfectly well to real materials and often to irreversible processes.

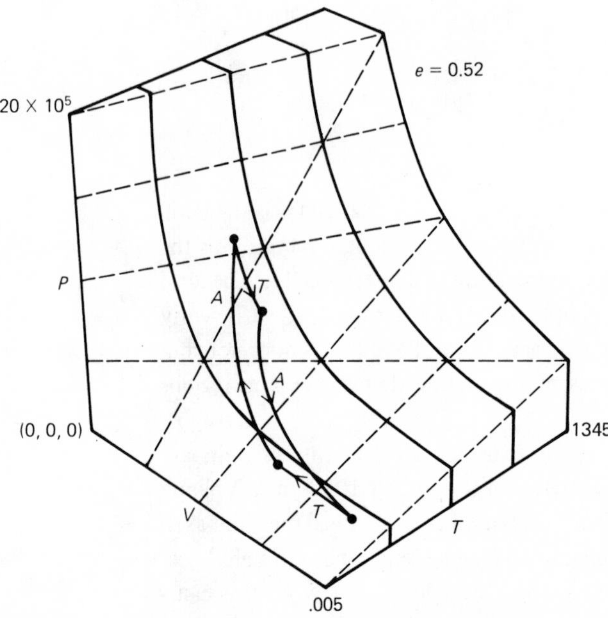

Figure 2.2 A Carnot cycle performed on an ideal gas. A Carnot cycle is two isotherms (T) and two adiabatic processes (A). Efficiency e is the work performed, W, divided by the heat absorbed, Q.

2.3 Illustrative Processes

Figure 2.2 shows a Carnot cycle performed on an ideal gas. The *PVT* surface is that of Figure 2.1. A Carnot cycle consists of two isothermal and two adiabatic processes. (An adiabatic process is one for which no heat is transferred to or from the system; PV^{γ} = a constant, where $\gamma = C_p/C_v$ and C_p, C_v are the specific heats at constant pressure and volume.) The isotherms are labeled *T*; the adiabatics are labeled *A*. The efficiency $e = W/Q$ (work performed per unit heat consumed) is 0.52 for this particular process. This efficiency is a function only of the temperatures T_{upper} and T_{lower}. All reversible processes between these *T*'s have the same efficiency, and no irreversible process may have greater efficiency.

The computer approach to thermodynamic material provides an easy way to calculate the heat, work, change in internal energy, and change in entropy for any process. You simply move along the *PVT* contour for the process [using the *PVT* relationship to calculate *P* (or *V* or *T*) in terms of *V* and *T* (or *P* and *T* or *P* and *V*)] and at each step calculate the heat used (usually $\Delta Q = C\,\Delta T$, where *C* is the heat capacity) and the work done (usually as $\Delta W = P\,\Delta V$). The total heat *Q* supplied to the system is then $\Sigma\,\Delta Q$, and the total work done is $\Sigma\,\Delta W$. The change in internal energy *U* is calculated from the first law of thermodynamics:

$$\Delta U = Q - W$$

The change in entropy is calculated by adding up ΔS at each step using the definition

$$\Delta S = \frac{\Delta Q}{T} \quad \text{and} \quad S = \sum \Delta S$$

The program *PVT* [Programs 2.1(a), (b)] illustrates one way to do these calculations for isothermal, isochoric, isobaric, and adiabatic processes. The integrations are done in the simplest way; you can use Simpson's rule if you wish. The specific heat at constant volume C_V is 1.5 ($\frac{3}{2}$) for an ideal (monatomic) gas; the specific heat at constant pressure C_P is 2.5 ($\frac{5}{2}$). The ratio γ is $\frac{5}{3}$ or 1.667.

Figure 2.3 shows a Diesel cycle, named after the same man as is the Diesel engine. The Diesel cycle is one isobar, two adiabatics, and one isochor. The efficiency, formed by the ratio W/Q, is $e = 0.48$. The four corners have $(T, V, P) = (216, 0.004, 10^5)$, $(554, 0.001, 1.01 \times 10^6)$, $(1{,}000, 0.001, 1.85 \times 10^6)$ and $(396, 0.004, 1.84 \times 10^5)$.

Figure 2.4 shows an Otto cycle (two isochors and two adiabatics).

The computer method can be used on real materials. You must be sure to use the right *PVT* relations in the right phases. Liquids and solids are nearly incompressible and also have specific heats substantially different from gases. Unlike the ideal gas, all the parameters for a real substance usually depend on pressure and temperature. Many materials (such as water) have several different solid phases in addition to liquid and gas phases.

Program 2.1(a)

PVTBAS *Finds ΔW, ΔQ, ΔU, ΔS for isothermal, isochoric, isobaric and adiabatic processes*

```
100 READ R,N,C,G                        } Thermodynamic constants
110 DATA 8.314,.223,12.45,1.667         }
120 READ I,Y0,X0,X1   Process, initial and final states
130 DATA 1,275,.0045,.003,4,169952,.003,.001             } Carnot cycle
140 DATA 1,572.233,.001,.0015,4,707288,.0015,.0045      }
150 LET W=0
160 LET Q=0
170 LET S=0
180 LET D=(X1-X0)/1000   Step size for integration
190 IF I=2 THEN 330
200 IF I=3 THEN 440
210 IF I=4 THEN 570
220 LET Z0=N*R*Y0/X0      }
230 FOR J=1 TO 1000       }
240 LET V=X0+(J-.5)*D     }
250 LET P=N*R*Y0/V        }
260 LET W=W+P*D           } Isothermal
270 LET Q=Q+P*D           } process
280 LET S=S+P*D/Y0        } (I = 1)
290 NEXT J                }
300 LET Y1=Y0             }
310 LET Z1=N*R*Y1/X1      }
320 GOTO 680              }
330 LET Z0=N*R*X0/Y0      }
340 FOR J=1 TO 1000       }
350 LET T=X0+(J-.5)*D     }
360 LET P=N*R*T/Y0        }
370 LET Q=Q+N*C*D         } Isochoric
380 LET S=S+N*C*D/T       } process
390 LET W=0               } (I = 2)
400 NEXT J                }
410 LET Y1=Y0             }
420 LET Z1=N*R*X1/Y1      }
430 GOTO 680              }
440 LET Z0=Y0*X0/(R*N)                       }
450 LET T2=Z0                                }
460 FOR J=1 TO 1000                          }
470 LET V=X0+(J-.5)*D                        }
480 LET T=Y0*V/(R*N)                         }
490 LET Q=Q+N*C*G*(T-T2)                     } Isobaric
500 LET S=S+N*C*G*(T-T2)/(T-(T-T2)/2)        } process
510 LET W=W+Y0*D                             } (I = 3)
520 LET T2=T                                 }
530 NEXT J                                   }
540 LET Y1=Y0                                }
550 LET Z1=Y1*X1/(N*R)                       }
560 GOTO 680                                 }
570 LET K=Y0*X0↑G         }
580 LET Z0=Y0*X0/(R*N)    }
590 FOR J=1 TO 1000       }
600 LET V=X0+(J-.5)*D     }
610 LET P=K/V↑G           } Adiabatic
620 LET W=W+P*D           } process
630 LET Q=0               } (I = 4)
640 LET S=0               }
650 NEXT J                }
660 LET Y1=K/X1↑G         }
670 LET Z1=Y1*X1/(N*R)    }
680 PRINT I,Q,W,Q-W,S
690 PRINT "INITIAL PT.:";X0;Y0;Z0;"  FINAL PT.:";X1;Y1;Z1
700 GOTO 120   Return for next process
710 END
```

W ≡ ΔW
Q ≡ ΔQ
S ≡ ΔS
When used,
P ≡ pressure
V ≡ volume
T ≡ temperature

Program 2.1(b)

PVTFOR *Finds ΔW, ΔQ, ΔU, ΔS for isothermal, isochoric, isobaric and adiabatic processes*

```
      REAL N
      DATA R,N,C,G/8.314,.223,12.45,1.6671/   Thermodynamic constants
1     READ(5,100)I,Y0,X0,X1    Process, initial and final states
      IF (I-999) 5,90,90
5     W=0.
      Q=0.
      S=0.
      D=(X1-X0)/1000.   Step size for integration
      IF (I-2) 6,20,6
6     IF (I-3) 7,40,7
7     IF (I-4) 8,60,8
8     Z0=N*R*Y0/X0              }
      DO 10 J=1,1000            }
      V=X0+(J-.5)*D             }
      P=N*R*Y0/V                }  Isothermal
      W=W+P*D                   }  process
      Q=Q+P*D                   }  (I = 1)
10    S=S+P*D/Y0                }
      Y1=Y0                     }
      Z1=N*R*Y1/X1              }
      GOTO 80
20    Z0=N*R*X0/Y0              }
      DO 30 J=1,1000            }
      T=X0+(J-.5)*D             }
      P=N*R*T/Y0                }  Isochoric
      Q=Q+N*C*D                 }  process
      S=S+N*C*D/T               }  (I = 2)
30    W=0.                      }
      Y1=Y0                     }
      Z1=N*R*X1/Y1              }
      GOTO 80
40    Z0=Y0*X0/(R*N)                              }
      T2=Z0                                       }
      DO 50 J=1,1000                              }
      V=X0+(J-.5)*D                               }
      T=Y0*V/(R*N)                                }  Isobaric
      Q=Q+N*C*G*(T-T2)                            }  process
      S=S+N*C*G*(T-T2)/(T-(T-T2)/2)               }  (I = 3)
      W=W+Y0*D                                    }
50    T2=T                                        }
      Y1=Y0                                       }
      Z1=Y1*X1/(N*R)                              }
      GOTO 80
60    AK=Y0*X0**G               }
      Z0=Y0*X0/(R*N)            }
      DO 70 J=1,1000            }
      V=X0+(J-.5)*D             }  Adiabatic
      P=AK/V**G                 }  process
      W=W+P*D                   }  (I = 4)
      Q=0.                      }
70    S=0.                      }
      Y1=AK/X1**G               }
      Z1=Y1*X1/(N*R)            }
80    PRINT 101,X0,Y0,Z0        }
      Q0=Q-W                    }  Print
      PRINT 102,I,Q,W,Q0,S      }  group
      PRINT 101,X1,Y1,Z1        }
      GOTO 1  Return for next process
90    STOP
100   FORMAT(I5,3F15.10)
101   FORMAT(3X,3E12.4)
102   FORMAT(1X,I5,4E12.4)
      END
```

W ≡ ΔW
Q ≡ ΔQ
S ≡ ΔS
When used,
P ≡ pressure
V ≡ volume
T ≡ temperature

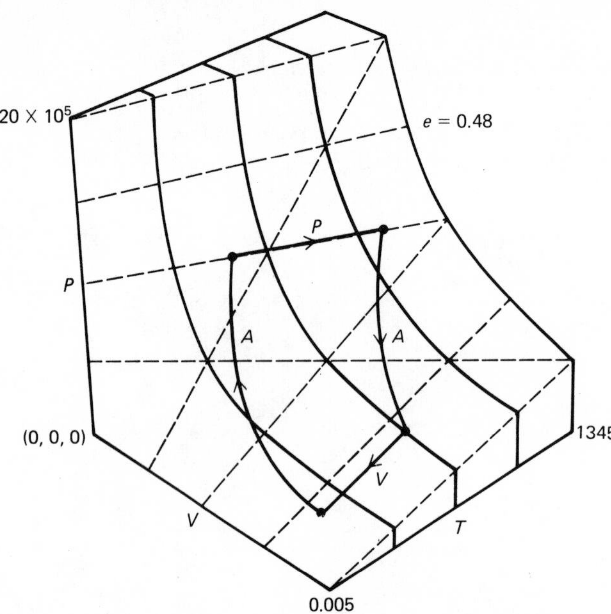

Figure 2.3 A Diesel cycle performed on an ideal gas. A Diesel cycle is two adiabatic processes (A), one isobar (P), and one isochor (V). The efficiency of the particular process shown is $e = 0.48$.

Problems

The following are representative of the exercises that have been done using the ideas developed in this chapter.

1. a. Find a Carnot cycle between STP and 77 K (liquid-nitrogen temperature). What is the efficiency of the process? Assume an ideal gas.
   b. Repeat part a for a van der Waals gas. Compare your results.

2. a. Find a Diesel cycle between STP and 77 K (liquid nitrogen temperature). What is the efficiency of the process? Assume an ideal gas.
   b. Repeat part a for a van der Waals gas. Compare your results.

3. a. Find an Otto cycle between STP and 77 K (liquid-nitrogen temperature). What is the efficiency of the process? Assume an ideal gas.
   b. Repeat part a for a van der Waals gas. Compare your results.

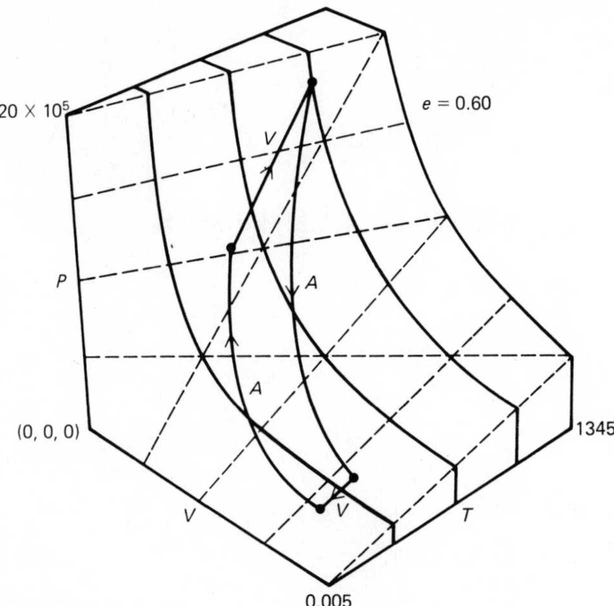

Figure 2.4 An Otto cycle performed on an ideal gas. An Otto cycle is two adiabatic processes (A) and two isochors (V).

4. a. Look up the *PVT* data for water.
   b. Enter the data into a file in the computer's memory.
   c. Write a version of the *PVT* program which gets its *PVT* data from your file (rather than by calculating from the ideal gas law.)
   d. Find a Carnot cycle using your true *PVT* relation for water.

3 statistical physics

3.1 The n-Body Gas

An interesting problem having to do with statistical processes that can be handled by a computer method is the n-body gas. n molecules (usually about a dozen) are started from particular initial positions and initial velocities. The motions of the (interacting) particles are displayed as time progresses. The most straightforward way to perform calculations is by use of a many-particle form of the $\mathbf{F} = m\mathbf{a}$ calculations discussed in the first chapter. Most gases are well approximated by two-particle interactions and elastic collisions with the container walls. A two-dimensional situation illustrates the effects almost as well as a three-dimensional one. The program NGAS [Programs 3.1(a), (b)] illustrates one possible program.

Two force laws have been used. The first is the hard-sphere approximation: the molecules of the gas are assumed to be spheres of some definite radius R and the only interaction between molecules occurs in elastic collisions. This situation approximates the ideal gas for fairly dilute situations ($nR^3 \ll L^3$, where the container is a cube of side L and there are n molecules). A second force law of interest is the Lennard-Jones 6–12 interaction. This interaction is one model of a real gas.

The temperature of the gas is set by the initial total kinetic energy (assuming that the initial situation has negligible particle interaction):

$$\tfrac{1}{2} m\overline{v^2} = \tfrac{3}{2} kT$$

Program 3.1(a)

NGASBAS *N-body gas*

```
100 DIM X(100),Y(100),U(100),V(100)
110 LET D=.01     Time step, Δt
120LET L=10     Length of box
130 LET T1=5     Print time
140 PRINT "# OF MOLECULES";
150 INPUT N
160 FOR I=1 TO N                             }
170 LET X(I)=(-1)↑INT(10*RND)*RND*L          } Initial
180 LET Y(I)=(-1)↑INT(10*RND)*RND*L          } positions
190 LET U(I)=(-1)↑INT(10*RND)*RND*L/10       } and
200 LET V(I)=(-1)↑INT(10*RND)*RND*L/10       } velocities
210 NEXT I                                   }
220 LET T0=0
230 LET T=0
240 LET D1=D/2     Half-time step, ½Δt
250 FOR I=1 TO N     Step through bodies
260 LET X0=X(I)
270 LET Y0=Y(I)
280 LET U=0
290 LET V=0
300 FOR J=I+1 TO N     Step through other interacting bodies
310 LET X2=X(J)-X0     }
320 LET Y2=Y(J)-Y0     } r from ith to jth bodies
330 LET R2=X2*X2+Y2*Y2 }
340 IF R2>16*.04 THEN 430     If separated far enough, do not interact
350 LET R6=R2*R2*R2                             }
360 LET F5=36E-12/(R6*R5*R2)-6E-6/(R6*R2)       } F of i on j
370 LET F1=F6*X2     Fx/m                       } (a 6-12 potential)
380 LET F2=F5*Y2     Fy/m                       }
390 LET U(J)=U(J)+F1*D1 }
400 LET V(J)=V(J)+F2*D1 } New j velocity
410 LET U=U-F1*D1 }
420 LET V=V-F2*D1 } New i velocity (temporary)
430 NEXT J
440 LET U(I)=U(I)+U }
450 LET V(I)=V(I)+V } New i velocity
460 NEXT I
470 FOR I=1 TO N     Step through bodies
480 LET X=X(I)+U(I)*D }
490 LET Y=Y(I)+V(I)*D } New position (temporary)
500 IF (X+L)*(X-L)<=0 THEN 590     If inside box, leave alone
510 LET S1=SGN(X)                                   }
520 LET X0=X(I)                                     }
530 LET X3=S1*L                                     } If outside
540 LET U=-U(I)                                     } box, reflect
550 LET D3=3*(X3-X0)/(X-X0)                         } off wall
560 LET X=X3+U*(D-D3)                               } elastically
570 LET U(I)=U                                      }
580 LET P1=P1+2*ABS(U)     Momentum for pressure    }
590 IF (Y+L)*(Y-L)<=0 THEN 680     If inside box, leave alone
600 LET S1=SGN(Y)                                   }
610 LET Y0=Y(I)                                     }
620 LET Y3=S1*L                                     } If outside
630 LET V=-V(I)                                     } box, reflect
640 LET D3=D*(Y3-Y0)/(Y-Y0)                         } off wall
650 LET Y=Y3+V*(D-D3)                               } elastically
660 LET V(I)=V                                      }
670 LET P1=P1+2*ABS(V)     Momentum for pressure    }
680 LET X(I)=X
690 LET Y(I)=Y     New position
700 NEXT I
710 LET T=T+D     New time
720 IF (T-T0)>=T1 THEN 750     Print test
730 LET D1=D     Remove initial half-step
740 GOTO 250
750 LET E1=0          }
760 FOR I=1 TO N      }
770 LET U=U(I)        }
780 LET V=V(I)        } Find total energy
790 LET E1=E1+U*U+V*V }
800 NEXT I            }
810 PRINT T,P1/(4*L*T1),E1
820 FOR I=1 TO N                 } Print positions
830 PRINT X(I),Y(I),U(I),V(I)    } and velocities
840 NEXT I
850 END
```

Program 3.1(b)

```
NGASFOR   N-body gas

        DIMENSION XMOL(100),YMOL(100),VXMOL(100),VYMOL(100)
        D=.01     Time step, Δt
        AL=10.    Length of box
        T1=5.     Print time
        DATA N/10/    Number of bodies
        AN=N
        DO 5 I=1,N                        } Initial positions
        XMOL(I)=(I-1)*2.*AL/AN-AL         } and velocities
        YMOL(I)=AL-(I-1)*2.*AL/AN         }
        VXMOL(I)=2.*AL/10.                }
5       VYMOL(I)=-2.*AL/20.               }
        T0=0.
        T=0.
        D1=D/2.    Initial half step, ½Δt
7       DO 25 I=1,N    Step through bodies
        X0=XMOL(I)
        Y0=YMOL(I)
        VX=0.
        VY=0.
        DO 20 J=I+1,N    Step through interacting bodies
        X2=XMOL(J)-X0       }
        Y2=YMOL(J)-Y0       } r from ith to jth body
        R2=X2*X2+Y2*Y2      }
        IF (R2-16.*.04) 10,20,20   If far enough separated, do not interact
10      R6=R2*R2*R2                                }
        FR=36.E-12/(R6*R6*R2)-6.E-6/(R6*R2)         } Force on jth body
        FX=FR*X2     F_x/m                          } due to ith body
        FY=FR*Y2     F_y/m                          }
        VXMOL(J)=VXMOL(J)+FX*D1   }
        VYMOL(J)=VYMOL(J)+FY*D1   } New jth velocity
        VX=VX-FX*D1    }
        VY=VY-FY*D1    } New ith velocity (temporary)
20      CONTINUE
        VXMOL(I)=VXMOL(I)+VX   }
25      VYMOL(I)=VYMOL(I)+VY   } New ith velocity
        DO 40 I=1,11       Step through bodies
        X=XMOL(I)+VYMOL(I)*D   }
        Y=YMOL(I)+VYMOL(I)*D   } New position (temporary)
        IF ((X+AL)*(X-AL)) 30,30,27   If inside box, leave alone
27      X0=XMOL(I)                   }
        X3=SIGN(S1*AL,X)             }
        VX=-VXMOL(I)                 } If outside box,
        D3=D*(X3-X0)/(X-X0)          } reflect off
        X=X3+VX*(D-D3)               } wall elastically
        VXMOL(I)=VX                  }
        P1=P1+2.*ABS(VX)  Momentum change for pressure
30      IF ((Y+AL)*(Y-AL)) 35,32,32   If inside box, leave alone
32      Y0=YMOL(I)                   }
        Y3=SIGN(S1*AL,Y)             }
        VY=-VYMOL(I)                 } If outside box,
        D3=D*(Y3-Y0)/(Y-Y0)          } reflect off
        Y=Y3+VY*(D-D3)               } wall elastically
        VYMOL(I)=VY                  }
        P1=P1+2.*ABS(VY)   Momentum change for pressure
35      XMOL(I)=X       New position
40      YMOL(I)=Y
        T=T+D    New time
        D1=D     Remove initial half-step
        IF (T-T1) 7,50,50     Print test
50      EN=0.
        DO 45 I=1,N            }
        VX=VXMOL(I)            } Find total
        VY=VYMOL(I)            } energy
45      EN=EN+VX*VX+VY*VY      }
        PRES=P1/(4.*AL*T1)   Pressure
        PRINT 101,T,PRES,EN
        DO 80 I=1,N                                        } Print positions
80      PRINT 100,XMOL(I),YMOL(I),VXMOL(I),VYMOL(I)        } and velocities
        STOP
100     FORMAT(1X,4E12.4)
101     FORMAT(1X,2HT=,F10.3,5H    P=,E12.4,5H    E=,E12.4)
        END
```

where m is the mass of each molecule, $\overline{v^2}$ is the mean-square velocity, and k is Boltzmann's constant, 1.38×10^{-23} J/deg. Since all the interactions conserve energy, the temperature (in most cases) remains constant as the speed and velocity distributions randomize toward Boltzmann distributions.

The interesting statistical phenomenon in the n-body gas is the development of a distribution of velocities. You can find the PVT relation by examining the same number of particles in several different-sized containers, each for several temperatures. The pressure is calculated from the average momentum change of the particles hitting the walls.

The difficulty of these calculations is the length of execution time; the computer time required is very great. Choose a method that minimizes the computer time whenever possible. It is often useful to calculate forces (for the 6–12 case, in particular) for small interparticle separation only and even then to use fourth-order Runge-Kutta methods or sophisticated predictor methods. To use the program NBODY on many computer systems, you should make the changes so that the calculations do not take too much computer time.

3.2 Statistical Averages and the Computer

All systems in thermal equilibrium have a well-defined distribution of velocities of the particles. One distribution function is the Boltzmann function:

$$f(\mathbf{r}, \mathbf{v})\, d\mathbf{r}\, d\mathbf{v} = A'' \exp\left[-\frac{E(\mathbf{r}, \mathbf{v})}{kT}\right] dx\, dy\, dz\, dv_x\, dv_y\, dv_z$$

where A'' is a normalization constant (a number to make $\iint f\, d\mathbf{r}\, d\mathbf{v} = 1$) and $E(\mathbf{r}, \mathbf{v})$ is the total energy of a particle at $\mathbf{r}$ with velocity $\mathbf{v}$. The distribution function f depends on both position ($\mathbf{r}$) and velocity ($\mathbf{v}$).

When there are no interactions between particles, the distribution simplifies to the Maxwell distribution. The total energy is the kinetic energy $\frac{1}{2}m(v_x^2 + v_y^2 + v_z^2)$:

$$f(v_x, v_y, v_z) = A' \exp\left[-\frac{m(v_x^2 + v_y^2 + v_z^2)}{2kT}\right]$$

The form of the Maxwell distribution is $f(v_x, v_y, v_z)$ integrated over all directions of v. The result is the Maxwell distribution of speeds:

$$f(v) = Av^2 \exp\left(-\frac{mv^2}{2kT}\right)$$

where

$$A = 4\pi \left(\frac{m}{2\pi kT}\right)^{3/2}$$

This distribution function f depends only on the speed $v = |\mathbf{v}|$.

The use of these distributions is usually limited to discussions of the average speed. Sometimes, the most probable speed and the rms (root-mean-square) speed of the particles are discussed. Once in a great while, there is a discussion of the law of atmosphere (wherein the gravitational potential energy of a particle, mgh, is included).

With the computer, interesting properties and systems can be treated using numerical integration. For example, the number of particles with speeds less than some v_0 can be integrated numerically; the analytic result contains the error function erf x, which is probably new to you and must be calculated numerically anyway. You can find the thermal average of any quantity, or the interparticle potential energy can be included. You can calculate the probability of finding a certain separation between particles. You can also find the fraction of the particles with kinetic energy greater than some minimum KE_0. Most chemical reactions demand some minimum available energy in order to start. Only those particles with larger kinetic energy are available for the reaction. All these statistical calculations are possible with numerical integration and the computer. The programs include a program to print out values of the Maxwell distribution of speeds [DSTFN, Programs 3.2(a), (b)], a program to find the probability of any molecule having a speed less than v_0 [DSTRFN, Programs 3.3(a), (b)], and a program to find the fraction of molecules having kinetic energy greater than a given energy [EMOLS, Programs 3.4(a), (b)]. All three programs use the Maxwell distribution of speeds for a noninteracting gas. (The last two programs use Simpson's rule for integration.)

Figure 3.1 shows the Maxwell speed distribution for a gas of hydrogen atoms at T = 300 K, 500 K, 1000 K, and 1500 K. Three hundred degrees Kelvin is room temperature; 500 K (200 °C) is about where solder melts; 1000 K (700 °C) is where

Program 3.2(a)

```
DSTFNBA   Distribution function evaluation

100 DEF FNM(V)=A*V*V*EXP(-M*V*V/(2*K*T))   Maxwell distribution function
110 LET M=1.67E-27   Mass of hydrogen
120 LET K=1.38E-23   Boltzmann's constant
130 LET T=300   Temperature
140 LET A=4*3.14159*(M/(2*3.14159*K*T))↑1.5   Coefficient
150 LET V0=SQR(2*K*T/M)
160 PRINT "V(M/S)","ENERGY","PROBABILITY"
170 FOR V=0 TO 3*V0 STEP V0/10   }
180 PRINT V,.5*M*V*V,FNM(V)      } Step through speeds
190 NEXT V                       }
200 END
```

Program 3.2(b)

```
DSTFNFOR      Distribution function evaluation

             FMAX(V)=A*V*V(EXP*-AM*V*V/(2.*AK*T))  Maxwell distribution
             AM=1.67E-27  Mass of hydrogen
             AK=1.38E-23  Boltzmann's constant
             T=300.  Temperature
             A=4.*3.14159265*SQRT((AM/(2.*3.14159265*AK*T))**3)  Coefficient
             V0=SQRT(2.*AK*T/AM)
             DO 10 I=1,31   Step through speeds
             V=(I-1)*V0/10.
             E=.5*AM*V*V
             F=FMAX(V)
  10         PRINT 100,V,E,F
             STOP
 100         FORMAT(1X,3E13.5)
             END
```

Program 3.3(a)

```
DSTRFNBA   Integration of distribution function

100 DEF FNM(V)=A*V*V*EXP(-M*V*V/(2*K*T))  Boltzmann distribution
110 LET M=1.67E-27   Mass of hydrogen
120 LET K=1.38E-23   Boltzmann's constant
130 LET T=300     Temperature
140 LET A=4*3.14159*(M/(2*3.14159*K*T))↑1.5      Coefficient
150 LET V0=SQR(2*K*T/M)   Upper limit of integration
160 LET D=SQR(2*K*T/M)/128      Integration step
170 FOR V=0 TO V0-D STEP D                               }
180 LET S1=S1+(FNM(V)+FNM(V+D))*D/2    Trapezoid sum    } Integration
190 LET S2=S2+FNM(V+D/2)*D  Midpoint sum                 }
200 NEXT V
210 PRINT S1/3+2*S2/3      Simpson's rule
220 END
```

Program 3.3(b)

```
DSTRFNFO    Integration of distribution function

             FMAX(V)=A*V*V*EXP(-AM*V*V/(2.*AK*T))  Boltzmann distribution
             AM=1.67E-27    Mass of hydrogen
             AK=1.38E-23    Boltzmann's constant
             T=300.  Temperature
             A=4.*3.14159*SQRT((AM/(2.*3.14159*AK*T))**3)   Coefficient
             V0=SQRT(2.*AK*T/AM)   Upper limit of integration
             D=SQRT(2.*AK*T/AM)/128.    Integration step
             S1=0.
             S2=0.
             N=V0/D     Number of intervals in integration sum
             DO 10 I=1,N
             V=(I-1)*D     Speed
             S1=S1+(FMAX(V)+FMAX(V+D))*D/2.     Trapezoidal sum
  10         S2=S2+FMAX(V+D/2.)*D     Midpoint sum
             S=S1/3.+2.*S2/3.     Simpson's rule
             PRINT 100,T,V0,S
             STOP
 100         FORMAT(1X,3E13.5)
             END
```

Program 3.4(a)

EMOLSBAS *Fraction of molecules with energy above given energy*

```
100 DEF FNM(V)=A*V*V*EXP(-M*V*V/(2*K*T))   Boltzmann distribution
110 LET M=1.67E-27   Mass of hydrogen atom
120 LET K=1.38E-23   Boltzmann's constant
130 LET T=300   Temperature
140 LET E0=K*T   Given energy
150 LET V0=SQR(2*E0/M)   Speed at given energy
160 LET V9=5*SQR(2*K*T/M)   Speed in far tail of distribution
170 LET A=4*3.14159*(M/(2*3.14159*K*T))↑(3/2)   Coefficient
180 LET D=SQR(2*K*T/M)/100   Step size for integration
190 LET S1=0
200 LET S2=0
210 FOR V=V0 TO V9 STEP D                              ⎫
220 LET S1=S1+(FNM(V)+FNM(V+D))*D/2   Trapezoidal sum  ⎬ Integration
230 LET S2=S2+FNM(V+D/2)*D   Midpoint sum              ⎭ loop
240 NEXT V
250 PRINT S1/3+2*S2/3   Simpson's rule
260 END
```

Program 3.4(b)

EMOLSFOR *Fraction of molecules with energy above given energy*

```
      FMAX(V)=A*V*V*EXP(-AM*V*V/(2.*AK*T))
      AM=1.67E-27   Mass of hydrogen atom
      AK=1.38E-23   Boltzmann's constant
      T=300.   Temperature
      E0=AK*T   Given energy
      V0=SQRT(2.*E0/AM)   Speed at given energy
      VEND=5.*SQRT(2.*AK*T/AM)   Speed in far tail of distribution
      A=4.*3.14159265*SQRT((AM/(2.*3.14159265*AK*T))**3)   Coefficient
      D=SQRT(2.*AK*T/AM)/100.   Step size for integration
      N=(VEND-V0)/D   Number of intervals in integration
      S1=0.
      S2=0.
      DO 10 I=1,N+1                                         ⎫
      V=V0+((I-1)*D)                                        ⎬ Integration
      S1=S1+(FMAX(V)+FMAX(V+D))*D/2.   Trapezoidal sum      ⎭ loop
10    S2=S2+FMAX(V+D/2.)*D   Midpoint sum
      S=S1/3.+2.*S2/3.   Simpson's rule
      PRINT 100,S
      STOP
100   FORMAT(1X,E12.4)
      END
```

materials begin to glow a dull red because they are warm; 1500 K (1200 °C) is about the temperature of a tungsten filament in an incandescent light bulb (the bulb looks white because of the long tail of the energy distribution which extends through the visible wavelengths).

Figure 3.2 shows the 1000 K distribution in more detail, with *mp* the most probable speed; *av* the average speed; and *rms* the rms speed, $\sqrt{\overline{v^2}}$. The kinetic energy of an atom with the rms speed is $\frac{3}{2}kT$.

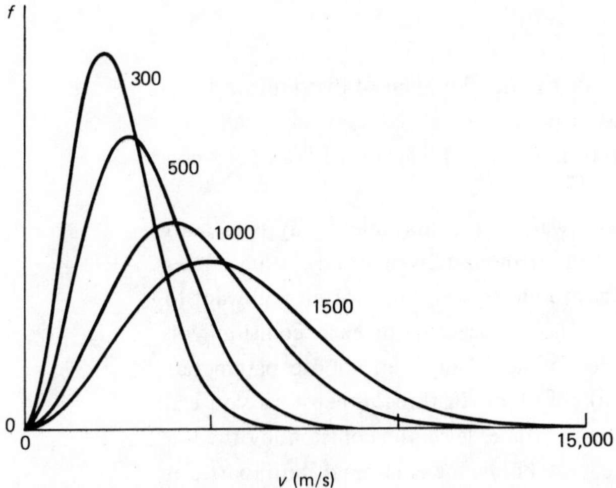

Figure 3.1 Maxwell speed distributions for temperatures of 300 K, 500 K, 1000 K, and 1500 K. Speeds between 0 and 15,000 m/s are displayed.

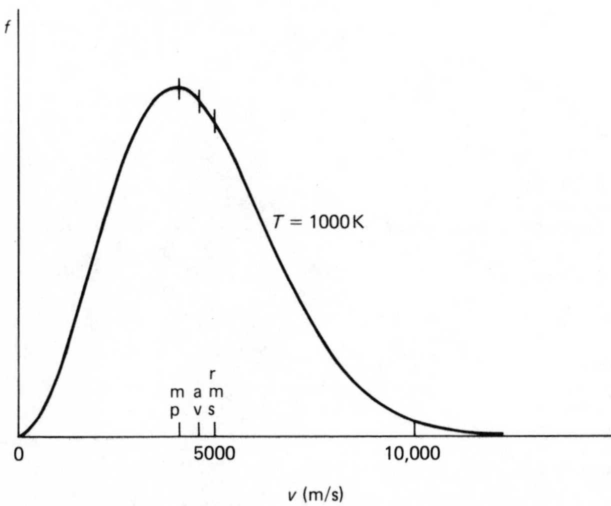

Figure 3.2 Maxwell speed distribution for 1000 K showing the most probable speed, the average speed, and the rms speed.

3.3 Gas with Random Collisions

Using a computer you can gain a feeling for the development of distribution functions by examining the development of a gas of point particles colliding elastically but randomly with each other. A program such as GAS3D [Programs 3.5(a), (b)] performs such calculations.

You can start the particles in any way you want. One convenient way is to start the particles with random energies in some range around a given average value. After choosing how many collisions you want the gas to undergo and how often you want to print out the distribution, you start the collision process. For each collision you randomly choose two particles which are to collide. You then choose parameters which will completely determine the (elastic) collision. In three dimensions you can choose (the cosines of) the angles at which the particles leave the collision; by the law of cosines and the fact that conservation of energy holds, the energies of both particles after the collision are then determined. (In one dimension you can choose either particle's new energy after the collision. Conservation of energy then gives the other particle's energy. Such a situation does not conserve momentum.) Using the com-

Program 3.5(a)

```
GAS3DBA    Gas with random collisions

100 DIM E(2000),D(26)
110 LET N=2000      Number of particles
120 LET E1=12.5/25    Energy in each ΔE
130 LET N0=1000     Number of collisions between prints
140 FOR I=1 TO N            } Initial energy
150 LET E(I)=1+2*RND        } and density
160 LET N1=1+INT(E(I)/E1)   } distribution
170 LET D(N1)=D(N1)+1       }
180 NEXT I
190 PRINT "ENERGY","NO. NEAR ENERGY"
200 FOR K=1 TO 5     Print 5 sets
210 FOR I=1 TO N0    Step through collisions
220 LET J0=1+N*RND   } Choose two
230 LET J1=1+N*RND   } particles randomly
240 LET E0=E(J0)+E(J1)       Total energy
250 LET N1=1+INT(E(J0)/E1)   } Remove first
260 LET D(N1)=D(N1)-1        } particle from density
270 LET N1=1+INT(E(J1)/E1)   } Remove second
280 LET D(N1)=D(N1)-1        } particle
290 LET C0=RND   } Choose collision's
300 LET C1=RND   } parameters randomly
310 LET E(J0)=E0/2+SQR(E0*E0-4*E(J0)*E(J1)*C0*C0)*C1/2   New first energy
320 LET E(J1)=E0-E(J0)    New second energy (conservation of energy)
330 LET N1=1+INT(E(J0)/E1)   } Add new first particle
340 LET D(N1)=D(N1)+1        } to density
350 LET N1=1+INT(E(J1)/E1)   } Add new second
360 LET D(N1)=D(N1)+1        } particle
370 NEXT I
380 FOR I=1 TO 26            } Print out
390 PRINT (I-.5)*E1,D(I)     } density of
400 NEXT I                   } particles
410 NEXT K
420 END
```

Program 3.5(b)

```
GAS3DFO    Gas with random collisions

      DIMENSION EN(2000),NDENS(26)
      NPART=2000      Number of particles
      DELTE=12.5/25.    ΔE for each density
      NCOLL=1000   Number of collisions between prints
      DO 10 I=1,NPART                 } Initial energy
      EN(I)=1.+2.*RANF(0.)            } and density
      N=1+INT(EN(I)/DELTE)            } distributions
10    NDENS(N)=NDENS(N)+1             }
      DO 40 K=1,5     Number of print-outs
      DO 50 I=1,NCOLL     Step through collisions
      J0=1+NPART*RANF(0.)    } Choose colliding
      J1=1+NPART*RANF(0.)    } particles randomly
      ETOT=EN(J0)+EN(J1)    Total energy of pair
      N=1+INT(EN(J0)/DELTE)     } Remove first particle
      NDENS(N)=NDENS(N)-1       } from density
      N=1+INT(EN(J1)/DELTE)     } Remove second particle
      NDENS(N)=NDENS(N)-1       } from density
      C0=RANF(0.)   } Choose scattering
      C1=RANF(0.)   } parameters randomly
      EN(J0)=ETOT/2.+SQRT(ETOT**2-4.*EN(J0)*EN(J1)*C0*C0)*C1/2.  New energy (first particle)
      EN(J1)=ETOT-EN(J0)   New energy (second particle) cons. of energy
      N=1+INT(EN(J0)/DELTE)    } Add new first particle
      NDENS(N)=NDENS(N)+1      } to density
      N=1+INT(EN(J1)/DELTE)    } Add second particle
50    NDENS(N)=NDENS(N)+1      } to density
      DO 40 I=1,26                }
      AI=I                        } Print out
      E=(AI-.5)*DELTE             } density of particles
40    PRINT 100,E,NDENS(I)        }
100   FORMAT(1X,F10.2,I8)
      STOP
      END
```

puter's random number generator, you assign (randomly) a new value to the chosen parameter.

You perform these random collisions the chosen number of times. It is convenient to keep track of the distribution function (the number of particles having various energies) as you go along. This is most easily done by defining a density vector whose elements represent the numbers of particles having energies between 0 and ΔE, ΔE and $2\Delta E$, $2\Delta E$ and $3\Delta E$, and so forth. To examine the distribution function at any time, you print out the distribution vector (or plot the vector elements versus energy).

As the number of collisions increases, the distribution will approach a Maxwell-Boltzmann distribution. One extension of these ideas examines the particles after a set number of collisions in phase space. You can watch the random collisions force the gas closer and closer to the well-known distribution function.

3.4 The Ising Model: Monte Carlo Calculations

Statistical processes can often be simulated using a random number generator. The Ising model of a (two-dimensional) lattice of spins is an illustration. Consider a lattice of spins (a grid with a spin at each grid point) which point either up or down. Assume the lattice of spins is placed in a uniform magnetic field. Each spin has two possible energies for the two possible spin directions relative to the magnetic field. If, in addition, each spin feels the magnetic field from neighboring spins, a large number of possible energies result depending on how the neighboring spins are pointed (and, hence, how the fields from these neighboring spins are directed).

If the coupling between neighboring spins is strong enough, all the spins will line up spontaneously at sufficiently low temperatures. If the coupling between spins makes neighbors want to align parallel to each other, a ferromagnet results; if neighbors want to be antiparallel, an antiferromagnet results. At sufficiently high temperatures this coupling between spins is overcome by thermal agitation, so that the spins can not align even if they try.

For a Monte Carlo approach to this problem, proceed in the following way. Step through the spin lattice examining each spin in turn. Calculate the energy for the spin in its present direction and with its direction reversed. If the energy is lowered by flipping the spin, then flip the spin; if the energy is not lowered, then consider the Boltzmann factor, exp $(-\Delta E/kT)$, where ΔE is the energy difference between the two spin directions. If the energy is not lowered, the spin will flip statistically with a probability per unit time given by the Boltzmann factor. The Monte Carlo technique calls a random number (between 0 and 1) and compares that random number to the Boltzmann factor. If the random number is less than or equal to the Boltzmann factor, the computer flips the spin; if the random number is larger than the Boltzmann factor, the spin remains unchanged. This sequence of steps simulates the statistical process, provided that the random number generator is truly random. One program implementing this Monte Carlo technique for the Ising model is ISING [Programs 3.6(a), (b), (c)]. A more complete discussion of the Monte Carlo approach to the Ising model is given in D. P. Landau and R. Allen, *Am. J. Phys.,* **41**,394 (1973).

One by-product of this method is the ability to examine the spin lattice in detail. Not only can you calculate the average energy and average total spin (magnetization) of the lattice at a given temperature (averaged over many statistically possible states, that is, over many iterations across the grid) but you can also compute the susceptibility and specific heat of the system (which are related to fluctuations of the system around the average energy and magnetization). You can also print out the spin lattice after each iteration and examine a number of statistically equivalent possible states.

If the neighboring spin coupling is strong and the temperature not quite low enough to allow the completely ordered state, the system is trying hard to order itself but can not quite overcome thermal agitation. In this temperature range, fluctuations become large and short-range order becomes apparent. Regions of the lattice will be ordered even though the lattice can not order as a whole. The transition temperature for a ferromagnet is called the Curie temperature; that for antiferromagnet is called the Néel temperature. The regions near these transition temperatures (where fluctuations are large) are very important technologically.

Program 3.6(a) Block diagram for the strategy of an Ising model calculation. The computer steps through spin-lattice flipping spins when the total energy is lowered or when the Boltzmann factor exceeds a random number.

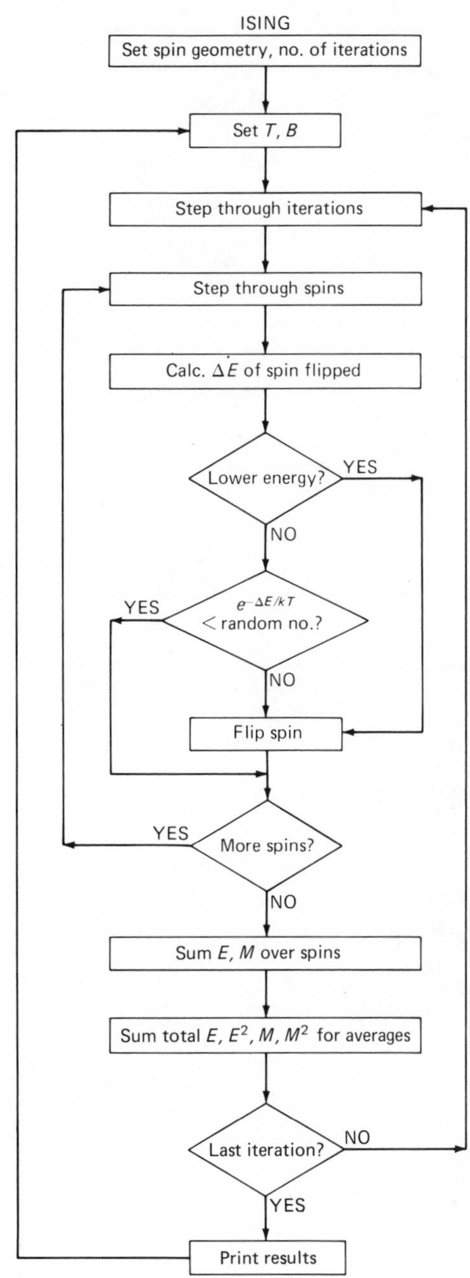

Program 3.6(b)

```
ISINGBAS    2D Ising model for interacting spins

100 DIM S(11,11),T(11,11)
110 PRINT "# OF SPINS ON A SIDE";
120 INPUT N0
130 PRINT "N.N  COUPLING CONST. (IN BOHR MAGNETONS)";
140 INPUT J0
150 PRINT "# OF CASES FOR ENSEMBLE AVERAGE";
160 INPUT N9
170 PRINT
180 PRINT "T(K), B(TESLA)";
190 INPUT T,B
200 FOR I=0 TO N0+1
210 FOR J=0 TO N0+1
220 LET S(I,J)=+1          Initialize
230 LET T(I,J)=+1          spins
240 NEXT J
250 NEXT I
260 LET N8=1     First time through
270 FOR K=1 TO N9    Step through iterations
280 FOR I=1 TO N0
290 FOR J=1 TO N0    Step through lattice
300 LET S0=T(I-1,J)+T(I+1,J)+T(I,J-1)+T(I,J+1)
310 LET E1=-9.274E-24*(B+J0*S0)*S(I,J)    Energy with spin up
320 LET E3=-E1    Energy with spin down
330 IF (E3-E1)<0 THEN 360    If gain energy, flip spin
340 LET E9=EXP(-(E3-E1)/(1.3806E-23*T))    e^-ΔE/kT
350 IF E9<RND THEN 370    Flip spin statistically
360 LET S(I,J)=-S(I,J)
370 NEXT J
380 NEXT I
390 FOR I=1 TO N0
400 LET T(N0+1,I)=S(1,I)
410 LET S(N0+1,I)=S(1,I)
420 LET T(0,I)=S(N0,I)
430 LET S(0,I)=S(N0,I)        Periodic
440 LET T(I,N0+1)=S(I,1)      boundary
450 LET S(I,N0+1)=S(I,1)      conditions
460 LET T(I,0)=S(I,N0)
470 LET S(I,0)=S(I,N0)
480 NEXT I
490 IF N8=1 THEN 630    First time through
500 LET E1=0
510 LET M1=0
520 FOR I=1 TO N0
530 FOR J=1 TO N0    Sum over spins
540 LET S0=S(I+1,J)+S(I-1,J)+S(I,J+1)+S(I,J-1)
550 LET E1=E1-.67174*(B+J0*S0)*S(I,J)/T    Energy
560 LET M1=M1+S(I,J)    Magnetization
570 NEXT J
580 NEXT I
590 LET E0=E0+E1    Sum for average energy
600 LET E2=E2+E1*E1    Sum for specific heat (fluctuations)
610 LET M0=M0+M1    Sum for average magnetization
620 LET M2=M2+M1*M1    Sum for susceptibility (fluctuations)
630 NEXT K
640 IF N8=2 THEN 710
650 LET E0=0
660 LET E2=0
670 LET M0=0
680 LET M2=0
690 LET N8=2    Second time through
700 GOTO 270
710 LET E0=E0/N9
720 LET E2=E2/N9    Average
730 LET M0=M0/N9    quantities
740 LET M2=M2/N9
750 PRINT "E AV.=";E0/(N0*N0);" KT/PART.=";E0*1.3806E-23*T
760 PRINT "SPEC. HEAT =";(E2-E0*E0)/(N0*N0);" KT UNITS/PART."
770 PRINT "M AV.=";M0/(N0*N0);" MAGN/PART =";M0;" MAGNETONS."
780 PRINT "SUSCEPT.=";.67174*(M2-M0*M0)/(N0*N0*T);"MAGN/TESLA-PART"
790 GOTO 170
800 END
```

Program 3.6(c)

```
ISINGFOR    2D Ising model for interacting spins

      REAL J0,MAG0,MAG1,MAG2,MAGAV
      DIMENSION SNEW(12,12),SOLD(12,12)
      DATA NSPIN,J0,NAV/10,.1,20/      Geometry
      ANSPIN=NSPIN                      of spins
      ANAV=NAV                          and averaging number
10    READ(5,100)T,B
      IF (T-999.) 5,90,90
5     DO 20 I=1,NSPIN+2
      DO 20 J=1,NSPIN+2        Initialize
      SNEW(I,J)=1.              spins
20    SOLD(I,J)=1.
      N8=1      First time through
22    DO 70 K=1,NAV     Step through iterations
      DO 30 I=2,NSPIN+1
      DO 30 J=2,NSPIN+1      Step through spins
      S0=SOLD(I-1,J)+SOLD(I+1,J)+SOLD(I,J-1)+SOLD(I,J+1)
      EN1=-9.274E-24*(B+J0*S0)*SNEW(I,J)     Energy with spin up
      E3=-EN1      Energy with spin down
      IF (E3-EN1) 28,25,25      If gain energy, flip spin
25    E9=EXP(-(E3-EN1)/(1.3806E-23*T))      e^-ΔE/kT
      IF (E9-RANF(0)) 30,28,28      Flip spin statistically
28    SNEW(I,J)=-SNEW(I,J)
30    CONTINUE
      DO 40 I=2,NSPIN+1
      SOLD(NSPIN+2,I)=SNEW(1,I)
      SNEW(NSPIN+2,I)=SNEW(1,I)
      SOLD(1,I)=SNEW(NSPIN,I)           Periodic
      SNEW(1,I)=SNEW(NSPIN,I)           boundary
      SOLD(I,NSPIN+2)=SNEW(I,2)         conditions
      SNEW(I,NSPIN+2)=SNEW(I,2)
      SOLD(I,1)=SNEW(I,NSPIN+1)
40    SNEW(I,1)=SNEW(I,NSPIN+1)
      IF (N8-1) 45,70,45      First time through
45    EN1=0.
      MAG1=0.
      DO 50 I=2,NSPIN+1
      DO 50 J=2,NSPIN+1      Sum over spins
      S0=SNEW(I+1,J)+SNEW(I-1,J)+SNEW(I,J+1)+SNEW(I,J-1)
      EN1=EN1-.67174*(B+J0*S0)*SNEW(I,J)/T      Energy of spins
50    MAG1=MAG1+SNEW(I,J)      Magnetization of spins
      EN0=EN0+EN1      Sum for average energy
      EN2=EN2+EN1*EN1      Sum for specific heat (fluctuations)
      MAG0=MAG0+MAG1      Sum for average magnetization
      MAG2=MAG2+MAG1*MAG1      Sum for susceptibility (fluctuations)
70    CONTINUE
      IF (N8-2) 80,85,80
80    EN0=0.
      EN2=0.
      MAG0=0.
      MAG2=0.
      N8=2      Second time through
      GOTO 22
85    EN0=EN0/ANAV
      EN2=EN2/ANAV        Average
      MAG0=MAG0/ANAV      quantities
      MAG2=MAG2/ANAV
      EAV=EN0/ANSPIN**2      Energy/spin
      SPHT=(EN2-EN0**2)/NSPIN**2      Specific heat
      MAGAV=MAG0/NSPIN**2      Magnetization/spin
      SUSC=.67174*(MAG2-MAG0**2)/(T*NSPIN**2)      Susceptibility
      PRINT 101,EAV,SPHT,MAGAV,SUSC
      GOTO 10
90    STOP
100   FORMAT(2F10.5)
101   FORMAT(1X,4E12.4)
      END
```

Problems

The following exercises represent ways the material in this chapter has been used with students.

1. a. Write a program to calculate the Maxwell speed distribution.
   b. Plot the Maxwell distribution for 10^{22} particles at 4 K, 77 K, 20 °C, 1000 °C and 10,000 °C.
   c. For each temperature from part b, determine how many particles have enough thermal energy to ionize a hydrogen atom.

2. a. Write a program to calculate the Maxwell speed distribution.
   b. Modify your program to treat the Boltzmann distribution for energies including both kinetic and gravitational (due to the earth alone).
   c. Using the general gravitational form for energy, show that the law of atmosphere (which assumes mgh for the potential energy) results when the particles are close to the surface of the earth.
   d. Investigate the distribution of particles far from the earth.

4 static electric fields

4.1 Introduction

Electrostatic fields are usually visualized by means of electric field lines and equipotential surfaces. Field lines are defined by two properties. (1) At any point, the tangent to the field line is parallel to the electric field at that point. (2) At any point, the number of field lines (crossing a unit cross section perpendicular to the lines) is proportional to the magnitude of the field at that point. Property 2 means that, as electric field lines get closer together, the electric field (and so the force on a test charge) gets larger. Property 1 furnishes an easy calculational way to draw field lines. Describing physical effects in terms of field-line patterns is basic to electricity and magnetism. After you get experience with field-line patterns for various charge distributions, you will better understand fields and how they are used.

Equipotential surfaces are surfaces in space on which you can move test charges without doing work; equipotentials are everywhere perpendicular to the electrostatic forces ($\mathbf{F} \cdot d\mathbf{l} = 0$). Thus, equipotentials are everywhere perpendicular to the electric field lines. You can follow an equipotential contour (the intersection of an equipotential surface and some chosen plane) by moving perpendicularly to the electric field at each point. This property provides an easy calculational way to draw equipotential contours.

4.2 Basic Formulas

The following formulas are discussed more fully in your text. If you do not recognize them, review your text. The formulas here merely identify the material discussed in this chapter. mks units will be used throughout.

1. $\mathbf{F} = \frac{1}{4\pi\epsilon_0} \frac{qq_1}{r^2} \hat{\mathbf{r}}$ *(Coulomb's law)*

   where $\mathbf{F}$ is the force on charge q due to charge q_1
   $\hat{\mathbf{r}}$ is a unit vector from charge q_1 to charge q

2. $\mathbf{F}_{\text{total}} = \mathbf{F}_1 + \mathbf{F}_2 + \mathbf{F}_3 + \cdots + \mathbf{F}_N = \sum_{i=1}^{N} \mathbf{F}_i = \sum_{i=1}^{N} \frac{qq_i}{4\pi\epsilon_0 r_i^2} \hat{\mathbf{r}}_i$

   $\mathbf{F}_{\text{total}}$ is the force on q due to all the charges $q_1, \ldots, q_N$.

3. The electric field

   $$\mathbf{E} = \lim_{q_0 \to 0} \left(\frac{\mathbf{F}}{q_0}\right)$$ *(Definition)*

   $\mathbf{E}$ is a vector field. It is a function whose values are vectors and which changes from point to point in space. $\mathbf{F}$ is the total force on q_0 at the observation point due to all charges (other than the test charge q_0).

4. Electrostatic field lines start on positive charges and end on negative charges. They start or end only on charges or at infinity.

5. The electric flux $\Phi = \int_S \mathbf{E} \cdot d\mathbf{S}$ for any surface S.

6. The electrostatic potential difference between points A and B is given by $V_{AB} = W_{AB}/q_0$. W_{AB} is the work done on q_0 by electric forces when the test charge q_0 moves from A to B.

   $$W_{AB} = \int_A^B \mathbf{F} \cdot d\mathbf{l}$$

7. $V = \frac{1}{4\pi\epsilon_0} \left(\frac{q}{r}\right)$

   for a point charge q relative to a zero potential at infinity.

8. All points on an equipotential surface are at the same electrostatic potential.

9. The electrostatic field lines are perpendicular to equipotential surfaces at every point.

4.3 Field-Line Calculations

Field lines can be calculated algorithmically, that is to say, in a step-by-step way. The process uses Property 1 of field lines mentioned above: tangents to field lines are everywhere parallel to the electric field. Let us limit ourselves to two-dimensional (or symmetric three-dimensional) situations (Figure 4.1). Consider the electric field at any point (x,y) in the plane formed by several coplanar point charges. The total electric field **E** is formed by adding up $(q/4\pi\epsilon_0)(\hat{\mathbf{r}}/r^2)$ for all the charges. Now consider stepping a short distance Δs along the field line from the point (x, y) to the point $(x + \Delta x, y + \Delta y)$. For small steps Δs, the triangle formed by the vectors Δx, Δy, and Δs is similar to the triangle formed by E_x, E_y, and $\sqrt{E_x^2 + E_y^2}$. So Δx and Δy are given by

$$\Delta x = \Delta s \left(\frac{E_x}{\sqrt{E_x^2 + E_y^2}} \right) \qquad \Delta y = \Delta s \left(\frac{E_y}{\sqrt{E_x^2 + E_y^2}} \right)$$

The new point on the field line is given by $(x + \Delta x, y + \Delta y)$ and the process starts again. In this way you march along the field line a step at a time.

The strategy used to follow an equipotential contour in the plane of the several point charges is nearly as simple. Since equipotentials are perpendicular to the electric field, at each step you move perpendicular to **E**. Perpendicular lines in a plane have slopes related by s and $-1/s$, so moving a distance Δs along an equipotential contour means moving distances

$$\Delta x = \Delta s \left(\frac{E_y}{\sqrt{E_x^2 + E_y^2}} \right) \qquad \Delta y = \Delta s \left(\frac{-E_x}{\sqrt{E_x^2 + E_y^2}} \right)$$

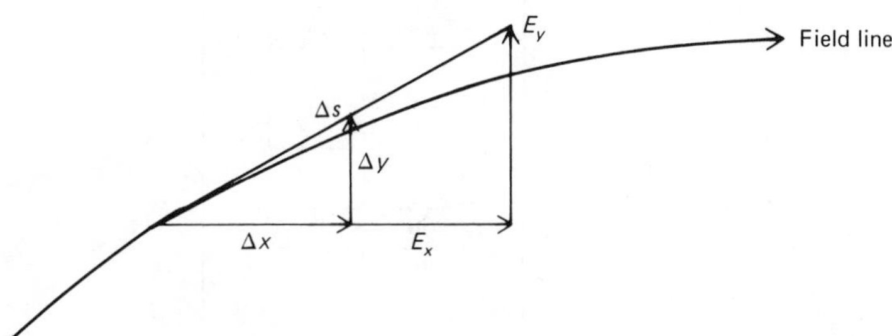

Figure 4.1 The triangles $(\Delta x, \Delta y, \Delta s)$ and $(E_x, E_y, |E|)$ are similar at any point on a field line.

parallel to the x and y axes. (The same equations result if at each step you rotate the x, y coordinate system by 90° with respect to the orientation used for field lines.)

The new point on the equipotential is $(x + \Delta x, y + \Delta y)$, and the process repeats.

A program which performs the algorithmic calculations to follow electric field lines and equipotential surfaces is EV [Programs 4.1(a), (b), (c)].

Using a program such as EV, you can plot pictures of electric field lines and equipotential surfaces for systems of N point charges. Other charge configurations (line charges, spherical charges, etc.) can also be used in the calculations. The method puts no constraint on the form of the charge distribution.

Figure 4.2 is a simple test of the method. The figure contains (1) the electric field lines starting at 45° intervals around a single point charge and (2) the equipotential surfaces for equally spaced potentials. The field lines are radial and the equipotentials are circular as expected.

In order to satisfy Property 2 for field lines, you should find a region of space where the field-line distribution is obvious. (For example, lines always start out radially near positive charges; twice as many lines start on +2 as on +1.) Once the lines are right somewhere, they are right everywhere.

Figure 4.3 is a slightly more complicated case of a dipole—a positive and negative charge of equal magnitude.

Program 4.1(a) Flow chart for a field-line and equipotential mapping strategy. After setting the initial values of parameters, the calculation walks along the field line or equipotential a step at a time. At each step the field at that point determines the components x and y of the next step.

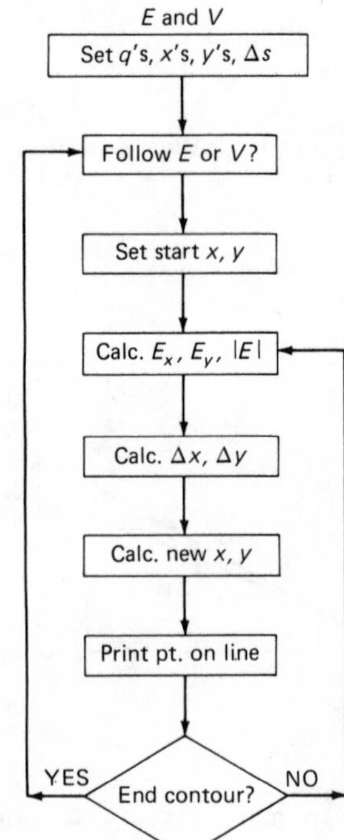

Program 4.1(b)

EVBAS *Electrostatic field lines and equipotentials*

```
100 DIM Q(10),X(10),Y(10)
110 PRINT "# OF POINT CHARGES";          Geometry
120 INPUT N0                              of source
130 FOR I=1 TO N0                         charges
140 PRINT " Q,X,Y FOR CHARGE #";I
150 INPUT Q(I),X(I),Y(I)
160 NEXT I
170 PRINT
180 PRINT "FIELD LINE (+1) OR EQUIPOTENTIAL (-1)";
190 INPUT E
200 PRINT "STARTING POINT (X,Y)";
210 INPUT X0,Y0
220 PRINT "X","Y"
230 LET X1=X0
240 LET Y1=Y0
250 LET D=0
260 LET F1=0
270 LET F2=0                         Initialization
280 LET F0=.1      Step size, Δs
290 LET F3=0
300 LET F4=0
310 FOR I=1 TO N0    Sum fields of all source charges
320 LET X9=X1+F1/2-X(I)              r from each
330 LET Y9=Y1+F2/2-Y(I)              source to field
340 LET R9=SQR(X9*X9+Y9*Y9)          point (half-stepped)
350 LET R9=R9*R9*R9      |r|3
360 LET F3=F3+Q(I)*X9/R9     Ex
370 LET F4=F4+Q(I)*Y9/R9     Ey
380 NEXT I
390 LET F5=SQR(F3*F3+F4*F4)     |E|
400 IF E=-1 THEN 440
410 LET F1=F0*F3/F5     Δx
420 LET F2=F0*F4/F5     Δy     Field line
430 GOTO 460
440 LET F1=F0*F4/F5     Δx
450 LET F2=-F0*F3/F5    Δy     Equipotential
460 LET X1=X1+F1     New x
470 LET Y1=Y1+F2     New y
480 LET D=D+F0
490 IF D<.5 THEN 520     Print
500 LET D=0              group
510 PRINT X1,Y1
520 IF ABS(X1-X0)+ABS(Y1-Y0)<.9*F0 THEN 170    Back to start?      Tests to
530 FOR I=1 TO N0                                                  end a
540 IF ABS(X1-X(I))+ABS(Y1-Y(I))<.9*F0 THEN 170  Near any          line
550 NEXT I                                       source charge?
560 IF ABS(X1)+ABS(Y1)>10 THEN 170    Far off page?
570 GOTO 290     Return for next Δs step
580 END
```

Program 4.1(c)

EVFOR *Electrostatic field lines and equipotentials*

```
      DIMENSION Q(10),XPOS(10),YPOS(10)
      READ(5,100)NUMQ                             Geometry of
      DO 10 I=1,NUMQ,1                            charges
10    READ(5,101) Q(I),XPOS(I),YPOS(I)
20    READ(5,101) CONTUR,XSTART,YSTART            Starting point
      X=XSTART
      Y=YSTART
      PDIST=0.
      DELTAX=0.
      DELTAY=0.                                   Initialization
      DELTAS=.1      Step size, Δs
30    XFIELD=0.
      YFIELD=0.
      DO 40 I=1,NUMQ,1      Sum fields of all source charges
      XTOQ=X+DELTAX/2.-XPOS(I)
      YTOQ=Y+DELTAY/2.-YPOS(I)                    r from each source
      RTOQ=SQRT(XTOQ**2+YTOQ**2)                  to field point (half-stepped)
      RCUBED=RTOQ**3     |r|^3
      XFIELD=XFIELD+Q(I)*XTOQ/RCUBED      Ex
40    YFIELD=YFIELD+Q(I)*YTOQ/RCUBED      Ey
      ABSFLD=SQRT(XFIELD**2+YFIELD**2)    |E|
      IF (CONTUR) 60,95,50
50    DELTAX=DELTAS*XFIELD/ABSFLD      Δx
      DELTAY=DELTAS*YFIELD/ABSFLD      Δy         Field line
      GOTO 70
60    DELTAX=-DELTAS*YFIELD/ABSFLD     Δx
      DELTAY=DELTAS*XFIELD/ABSFLD      Δy         Equipotential
70    X=X+DELTAX      New x
      Y=Y+DELTAY      New y
      PDIST=PDIST+DELTAS
      IF (PDIST.LT..5) GOTO 80                    Print
      PDIST=0.                                    group
      PRINT 102,X,Y                                              Tests to
                                                                 end a line
80    IF (ABS(X-XSTART)+ABS(Y-YSTART).LT..9*DELTAS) GOTO 20 Back to start?
      DO 90 I=1,NUMQ,1
90    IF (ABS(X-XPOS(I))+ABS(Y-YPOS(I)).LT..9*DELTAS) GOTO 20 Near any source charge?
      GOTO 30      Return for next Δs step
100   FORMAT(I2)
101   FORMAT(3F5.2)
102   FORMAT(1X,2E15.5)
95    STOP
      END
```

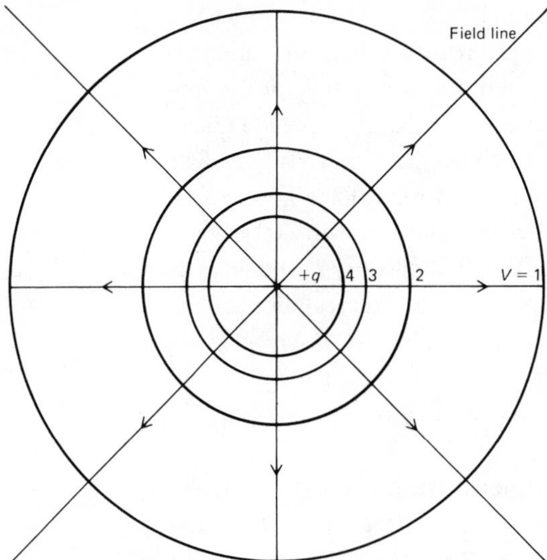

Figure 4.2 Field lines and equipotential contours for a single point charge. The field lines are chosen to start at 45° intervals around the charge; the equipotentials are for $V = 1$, 2, 3 and 4 (in normalized units).

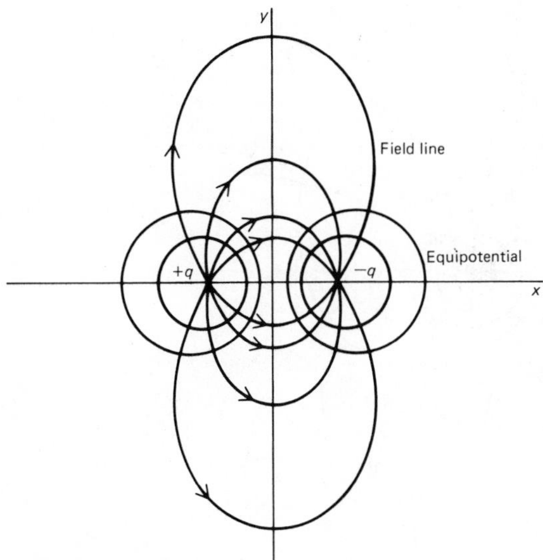

Figure 4.3 Field lines and equipotential contours for a dipole. The field lines start on the positive point charge and end on the equal negative point charge.

Figure 4.4 shows the electric field lines and equipotential contours in the plane of a two-dimensional quadrupole—two positive and two equal negative charges on alternating corners of a square. Notice how easy it is to get a qualitative feeling for the contours. The electric field lines start radially from the positive charges and end radially on the negative charges. The equipotentials start as circles near the charges. The equipotentials go to zero far from the charges and at the center of the quadrupole. The numerical method has a very hard time following contours with very sharp corners or following contours very near places where $\mathbf{E} = 0$. Steps, Δs, must always be small enough that the numerical solution will converge to the true solution of the differential equation as you integrate your way along the line. (Convergence is a problem in numerical methods; numerical methods are discussed in the Appendix.) Often the most straightforward way to test for convergence is to halve the step size and see if the solution changes significantly.

Figure 4.5 shows the pattern from a one-dimensional quadrupole—two dipoles head to head. Figure 4.6 shows the pattern of three equal charges on an equilateral triangle. Figure 4.7 is the pattern for four equal point charges on the corners of a

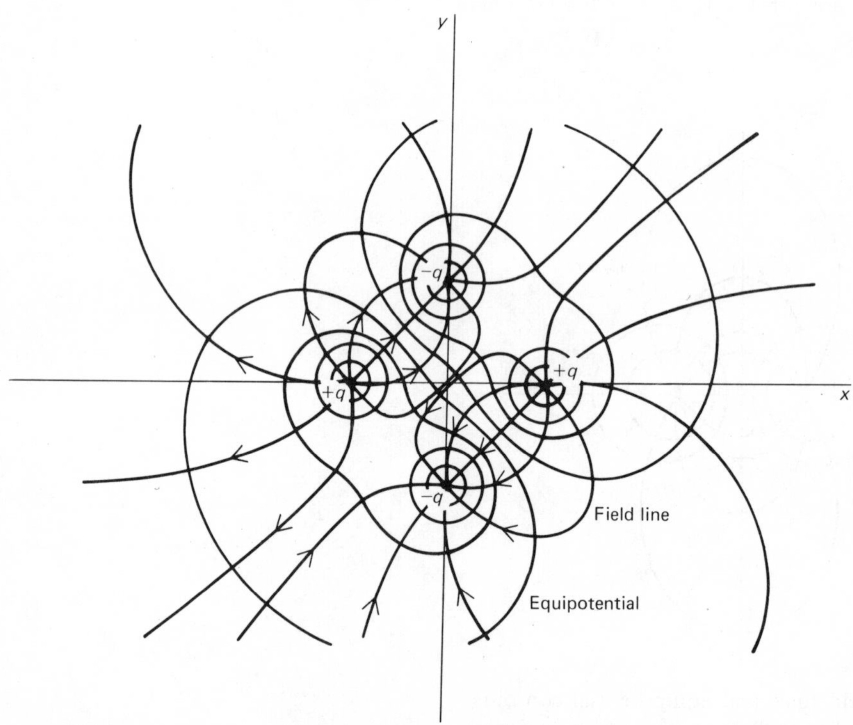

Figure 4.4 Field lines and equipotential contours for a two-dimensional quadrupole. Four point charges of equal magnitudes and alternating sign lie on the corners of a square.

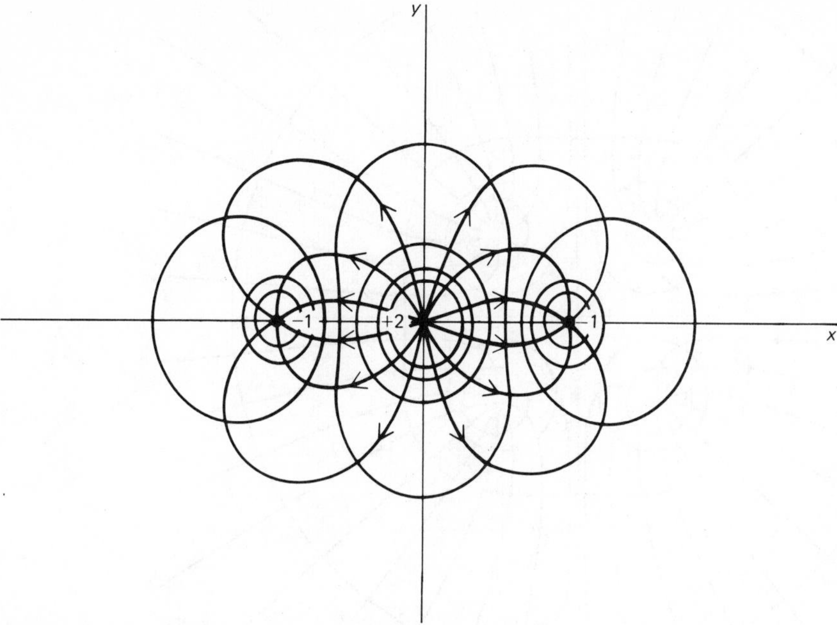

Figure 4.5 Field lines and equipotential contours for a one-dimensional quadrupole. There are point charges −1 on either side of a point charge +2.

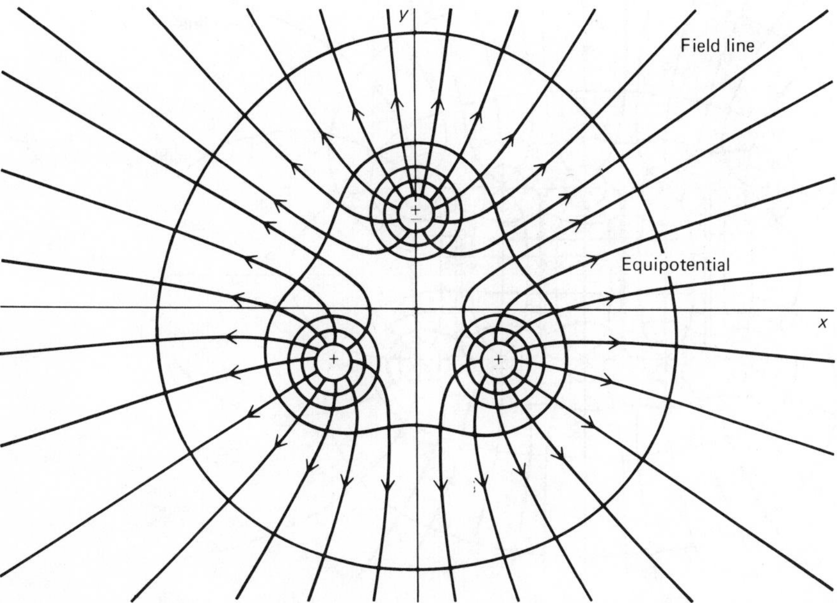

Figure 4.6 Field lines and equipotential contours for three equal point charges on the corners of an equilateral triangle.

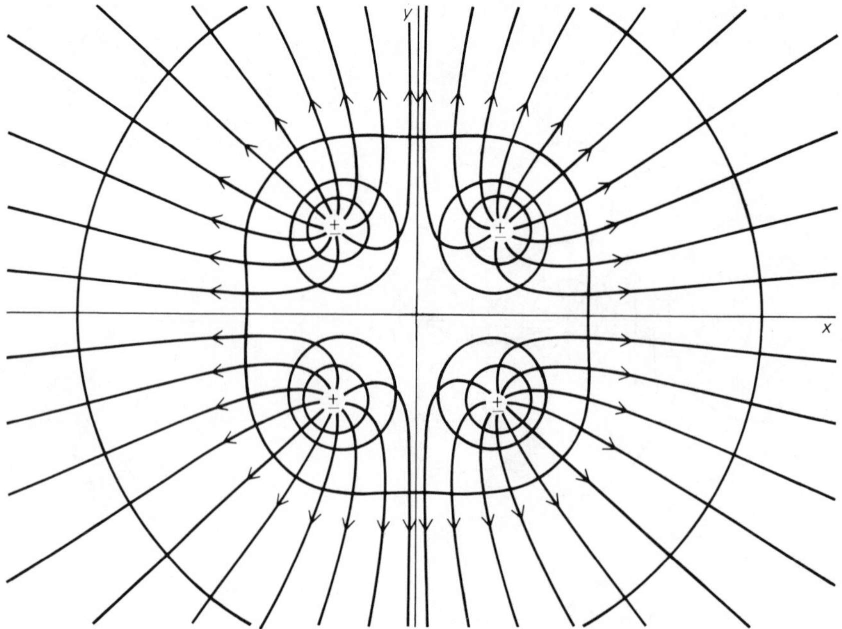

Figure 4.7 Field lines and equipotential contours for four equal point charges on the corners of a square.

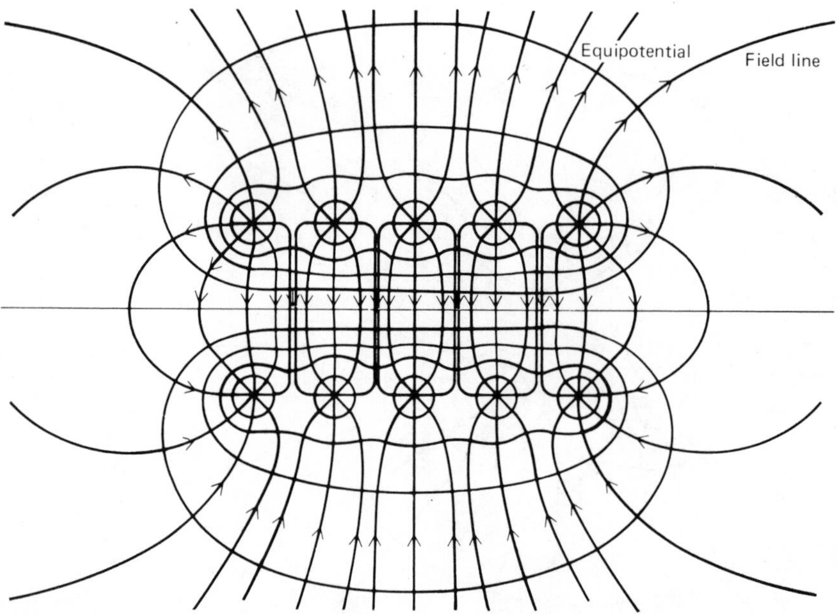

Figure 4.8 Model of a short capacitor. Field lines and equipotentials for a system of five positive and five negative point charges of equal magnitudes.

square. Figure 4.8 is an approximation to a short capacitor—five positive and five equal negative charges. In all these cases, you can sketch the field lines pretty well by deciding where they start and end. You can sketch the equipotentials by making them circular both near charges and far away. Check your intuition by sketching the field line and equipotential pattern for three equal point charges on the corners of an equilateral triangle. Compare your sketch with Figure 4.6. Although it is hard to guess at the right positions for equally spaced equipotentials, you should be able to reproduce the qualitative features of the pattern.

4.4 Charges between Conducting Plates

One of the interesting results of electrostatics states that the field pattern due to a point charge (or a set of point charges) near a (grounded) conducting plate (or a group of plates) is the same as that due to the original charge and a set of reflected image charges. If a single charge and a single flat conducting plate are present, there is only one image charge. The image charge has the same magnitude as, but an opposite sign to, that of the original charge. The image charge is the same distance behind the plate that the original charge is in front of the plate. When more plates are present, the image charges have image charges, and the situation becomes more complicated.

The computer can be used to plot the field lines and equipotentials after a calculation of the total field at each point. The computer can also be used to calculate the positions and values of all the image charges (even in very complicated cases).

QINCAP is a program which calculates field lines for a single point charge placed anywhere between two grounded, parallel, conducting plates [Programs 4.2(a) and (b)]. When the charge is midway between the plates, the positions of the image charges resemble a binomial series. When the charge is off center, the positions form a complicated series. The computer finds (and stores) the values and positions of the image charge iteratively; it calculates the images of image charges as it goes along.

Having stored the image charges, the computer calculates field lines (and equipotentials) by the iterative (step-by-step) method already discussed, assuming that the pattern is due to the original charge and all the images taken together.

4.5 Motion of Charges in E Fields

Many students have difficulty relating electric field lines and the trajectories of charged particles released in the electric field. With your experience (including your experience with $\mathbf{F} = m\mathbf{a}$ programs), you can demonstrate the difference between trajectories and field lines very easily. First, plot the field-line pattern for a fairly simple charge distribution—say the dipole with charges at (+1, 0) and (−1, 0). Now write a two-

Program 4.2(a)

QINCAPBA *Field lines and equipotentials for charge between two grounded plates*

```
100 DIM Q(20),Y(20)
110 PRINT "HEIGHT OF Q, SEPARATION OF PLATES?"
120 INPUT D,L
130 LET N0=20       Number of charges considered
140 LET Q(1)=+1   } Original
150 LET Y(1)=D    } charge
160 LET Q(2)=-1                       }
170 LET Y(2)=2*L-D                    }
180 LET Q(3)=-1                       }
190 LET Y(3)=-D                       }
200 FOR I=3 TO N0                     } Set
210 IF INT(I/2)<> I/2 THEN 250        } up
220 LET Y(I)=L+(L-Y(I-1))             } image
230 LET Q(I)=-Q(I-1)                  } charges
240 GOTO 270                          }
250 LET Y(I)=-Y(I-3)                  }
260 LET Q(I)=-Q(I-3)                  }
270 NEXT I                            }
280 PRINT "FIELD(+1) OR EQUI-V(-1), (X,Y) OF START?"
290 INPUT E, X1,Y1
300 PRINT "X","Y"
310 LET X0=X1
320 LET Y0=Y1
330 LET F1=0
340 LET F2=0
350 LET F0=.1      Step size, Δs
360 LET F3=0
370 LET F4=0
380 FOR I=1 TO N0      Sum over charges
390 LET X9=X1+F1/2                }  r from each
400 LET Y9=Y1+F2/2-Y(I)           }  charge to field point
410 LET R9=SQR(X9*X9+Y9*Y9)       }  (half-stepped)
420 LET R9=R9*R9*R9    |r|^3
430 LET F3=F3+Q(I)*X9/R9    E_x
440 LET F4=F4+Q(I)*Y9/R9    E_y
450 NEXT I
460 LET F5=SQR(F3*F3+F4*F4)    |E|
470 IF E=-1 THEN 510
480 LET F1= F0*F3/F5    Δx }
490 LET F2=F0*F4/F5     Δy } Field line
500 GOTO 530
510 LET F1=F0*F4/F5     Δx }
520 LET F2=-F0*F3/F5    Δy } Equipotential
530 LET X1=X1+F1     New x
540 LET Y1=Y1+F2     New y
550 PRINT X1,Y1
560 IF ABS(X1-X0)+ABS(Y1-Y0)< F0 THEN 580    Back to start?  } Tests to
570 IF Y1*(Y1-L) <0 THEN 360    Outside plates?              } end line
580 GOTO 280    Return for new contour
590 END
```

y = L
Q
D
y = 0

dimensional **F** = m**a** program using qE as the force at each point. Start your charge q at points such as (+1, +1) and (0, +1). Integrate **F** = m**a** to find the trajectories of the charge q. You will find that as soon as the charge gains momentum, it tends to leave the field line. The electric field only produces changes in momenta (by Newton's second law.)

Program 4.2(b)

QINCAPFO *Field lines and equipotentials for charge between two grounded plates*

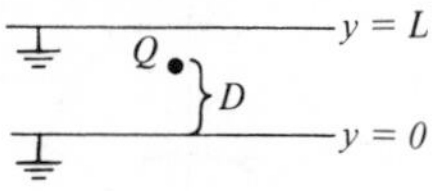

```
      DIMENSION Q(20),YQ(20)
      DATA HT,SEP,NUMQ/.5,1.,20/        Height of charge, separation
                                        of plates, number of charges considered
      Q(1)=1.       } Original charge
      YQ(1)=HT      }
      Q(2)=-1.                          }
      YQ(2)=2.*SEP-HT                   }
      Q(3)=-1.                          }
      YQ(3)=-HT                         }
      DO 10 I=4,NUMQ                    }
      AI=I                              }
      IF (I/2-AI/2.) 5,4,5              } Set up image charges
4     YQ(I)=SEP+(SEP-YQ(I-1))           }
      Q(I)=-Q(I-1)                      }
      GOTO 10                           }
5     YQ(I)=-YQ(I-3)                    }
      Q(I)=-Q(I-3)                      }
10    CONTINUE                          }
15    READ(5,100)LINE,XST,YST           Field line or equi-V and start point
      IF (LINE-999) 17,90,90
17    X=XST
      Y=YST
      DX=0.
      DY=0.
      DS=.1        Step size, Δs
20    EX=0.
      EY=0.
      DO 40 I=1,NUMQ        Sum over charges
      XTEM=X+DX/2.                      } r from each
      YTEM=Y+DY/2.-YQ(I)                } charge to field point
      RTEM=(XTEM**2+YTEM**2)**1.5       } (half-stepped)
      EX=EX+Q(I)*XTEM/RTEM      Ex
40    EY=EY+Q(I)*YTEM/RTEM      Ey
      E=SQRT(EX**2+EY**2)       |E|
      IF (LINE+1) 50,60,50
50    DX=DS*EX/E     Δx  } Field line
      DY=DS*EY/E     Δy  }
      GOTO 70
60    DX=DS*EY/E     Δx  } Equipotential
      DY=-DS*EX/E    Δy  }
70    X=X+DX       New x
      Y=Y+DY       New y
      PRINT 101,X,Y
      IF (ABS(X-XST)+ABS(Y-YST)-DS/2.) 15,15,80    Back to start?  } Tests to
80    IF (Y*(Y-SEP)) 20,15,15      Outside plates?                  } end contour
90    STOP
100   FORMAT(I5,2F10.5)
101   FORMAT(1X,2F10.4)
      END
```


Figure 4.9 shows some trajectories of positive point charges released in the field of a dipole.

Charges injected into the field of a single point charge undergo Keplerian orbits. Since charges come in two types, trajectories under a repulsive inverse-square law can also be studied. What might orbits around the charges of a dipole look like? Can you see how to use the program you have written to discover closed orbits in the field of a dipole?

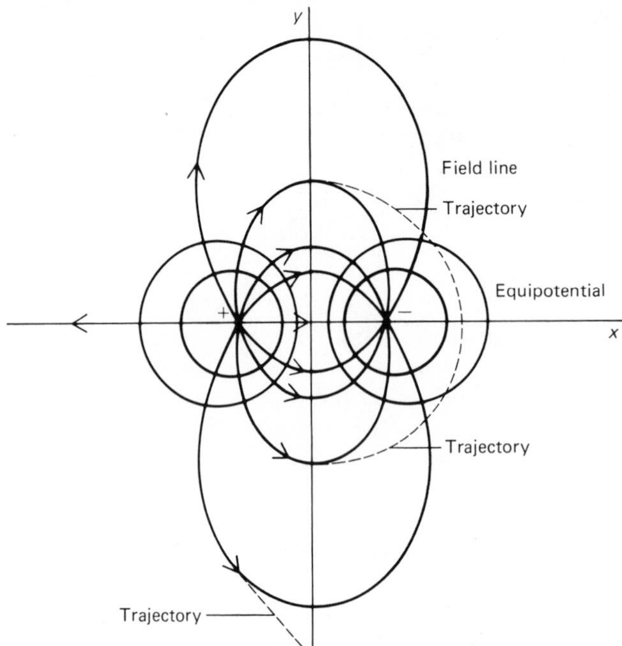

Figure 4.9 Trajectories (dotted) of point test charges in the field of a dipole.

Problems

A number of problems have been used with this material. A few of them, ranging over the whole field of electrostatics, follow.

Figure 4.10 Geometry of charges for Problem 1.

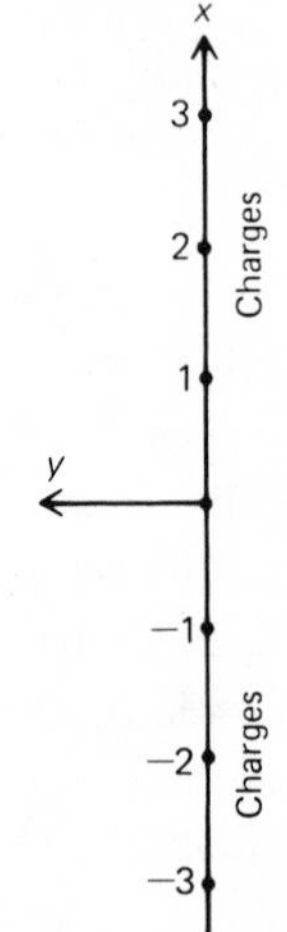

1. Calculate the force on a unit positive point charge at $(0, +y)$ due to 101 unit positive point charges placed symmetrically around the origin along the x axis (Figure 4.10). Separate the fixed charges by one space unit. Put one charge at the origin. Choose $y = 0.001$, 1, 10, 100, and 1000 space units. Show that for small y only the central fixed charge is important, whereas for intermediate values of y, the result can be approximated by an infinite uniformly charged rod. What happens when y gets very large?

2. Calculate the force on a test charge caused by 36 point charges placed at 10° intervals around a circle. Find the force at a number of points. Why is your answer for the force at the center of the circle not quite zero?

3. a. Approximate a uniformly charged square (one space unit on a side) carrying a total charge of 1 unit by a set of 100 point charges each of

0.01 unit. Calculate the electric field (in the plane of the charges) due to these charges at a number of points in space. How far away from the square must you be to have a good approximation? What happens when you are very far away?

b. Using your field calculations from 3a and a three-dimensional form of a field-line subroutine, calculate and plot several field lines for your approximation to the uniform square.

c. Using the field calculations from 3a, calculate and plot several equipotential contours in the plane of the square.

4. Using the field line and equipotential routine, calculate and plot field lines and equipotentials for several of the following charge distributions. For each case, sketch the lines as you think they will appear before you run the program.
   a. Point charges: +2 at (+1, 0); -1 at (0, 0)
   b. Point charges: +1 at the corners of a square; -4 at its center
   c. Point charges: +1 and -1 charges on alternating corners of an octagon (a two-dimensional octupole)
   d. Two small (with respect to the spacing) uniformly charged spherical volumes of opposite sign
   e. Four positive uniformly charged infinite line charges cutting the plane at the corners of a square

5. Using an $\mathbf{F} = m\mathbf{a}$ program, find several trajectories for (a) positive and (b) negative point charges in the field of a single positive fixed point charge. Why do elliptical closed orbits occur for the negative charge—the equipotentials for the system are all circular? (Remember that orbital motion has its total energy conserved.)

6. a. Write a program to calculate the flux through planes perpendicular to Cartesian coordinate axes. Find the flux through several such planes due to a point charge at the origin, a dipole of finite length centered at the origin, and a one-dimensional finite quadrupole centered at the origin. Keep in mind the field-line patterns for each charge distribution; make sure your results are sensible physically.
   b. Put your calculations together to find the total flux through the surface of a cube of size L centered on the origin. Show that Gauss' law holds for the charge distributions given in a and for several different cubes.

7. The electrostatic field is said to be conservative because the potential difference between any two points in the field is not dependent on the path from A to B. Write and run a program which calculates potential differences for piecewise straight paths with the pieces parallel to the Cartesian coordinate axes. Use the electric field due to four equal charges on the corners of a

square. Show that the potential difference is path-independent (at least for these paths). You cannot prove the independence for general paths using computers; proofs must be done analytically.

8. This problem demonstrates a way to follow equipotentials and field lines in a manner different from that discussed in the chapter.
   a. Place two equal positive charges at the points $(+1, 0)$ and $(-1, 0)$. Cover the square having corners $(\pm 2, \pm 2)$ with a grid of 100 points. Calculate the potential at every point on the grid. By looking at the numbers, follow several equipotential contours. By looking at the numbers to find the directions of steepest slope around each point, follow several field lines.
   b. Can you think of a computer approach for following equipotentials by storing the potentials on points of a grid and then hunting through the grid to follow the equipotential? Outline your approach. Such an approach is sometimes better than the one discussed in the text if many equipotentials are to be followed for one charge distribution. How would you follow field lines this way?

9. Using your own forms of the field line and equipotential program, find the field lines and equipotential contours for a couple of point-charge or line-charge distributions that you think might be interesting or pretty. Then by using an $\mathbf{F} = m\mathbf{a}$ program, show that test charges released in your fields will only very rarely follow field lines.

10. Think up two problems in which the computer can be used to make clearer some of the material you have just learned in the course text. Prove that these computer problems of yours can be solved by handing in programs and results.

5 vector calculus; laplace's and poisson's equations

5.1 Introduction

There are three derivatives of multivariable functions which involve vector relationships. All three of these vector calculus derivatives are used throughout physics, but the student usually comes across this vector calculus first in electricity and magnetism. The laws of electricity and magnetism are especially concisely stated in terms of these vector calculus derivatives; and analytical manipulations of the laws (such as the derivations of wave equations) are made especially easy when the laws are written with vector calculus derivatives.

The first of the vector calculus derivatives is called the gradient. The *gradient* is a vector that points in the direction of steepest increase of the multiple-variable function; the magnitude of the gradient is calculated by using partial differentiation. The original function itself is a scalar function: it assigns a number to every point in space, but it has no direction. The gradient, then, is a vector derivative of a scalar function.

Symbolically, the gradient is usually written as $\nabla f(x, y, z)$ for the scalar function f of three variables. (∇ is called "del.") By definition, the gradient of f is

$$\nabla f(x, y, z) = \hat{\mathbf{x}} \frac{\partial f}{\partial x} + \hat{\mathbf{y}} \frac{\partial f}{\partial y} + \hat{\mathbf{z}} \frac{\partial f}{\partial z}$$

where $\hat{\mathbf{x}}$, $\hat{\mathbf{y}}$, $\hat{\mathbf{z}}$ are unit vectors pointing in the x, y, z directions and $\partial f/\partial x$, $\partial f/\partial y$, $\partial f/\partial z$ are the first partial derivatives of the function f. (To take a partial derivative with respect to any variable, say x, you treat the other variables, in this case y and z, as constants and then take a derivative in the usual way.)

Another derivative in vector calculus is called the divergence. The *divergence* is a scalar related to derivatives of a vector function, so it is a scalar derivative of a vector function. A function which has vector values and depends on several variables is a vector function of a vector:

$$\mathbf{F}(x, y, z) = \mathbf{F}(\mathbf{r})$$

Such a function is called a *vector field* and has components and so forth just like any other vector. The forces you have been dealing with in physics, the gradient, and the electric field $\mathbf{E}$, are all examples of vector fields. By definition, using the common notation for the divergence,

$$\text{The divergence of } \mathbf{F}(x, y, z) \equiv \nabla \cdot \mathbf{F} \equiv \frac{\partial F_x}{\partial x} + \frac{\partial F_y}{\partial y} + \frac{\partial F_z}{\partial z}$$

The third kind of derivative of vector calculus is a vector derivative of a vector, or a *curl,* and its definition is

$$\text{Curl of } \mathbf{F}(x, y, z) \equiv \nabla \times \mathbf{F} = \hat{\mathbf{x}}\left(\frac{\partial F_z}{\partial y} - \frac{\partial F_y}{\partial z}\right) + \hat{\mathbf{y}}\left(\frac{\partial F_x}{\partial z} - \frac{\partial F_z}{\partial x}\right) + \hat{\mathbf{z}}\left(\frac{\partial F_y}{\partial x} - \frac{\partial F_x}{\partial y}\right)$$

Again, $\hat{\mathbf{x}}$, $\hat{\mathbf{y}}$, and $\hat{\mathbf{z}}$ are unit vectors, parallel to the x, y, z axes.

You can take higher derivatives in vector calculus, too. One of the most common and most useful is the Laplacian, named after the mathematician Laplace. The Laplacian is the divergence of the gradient of a scalar function. It is a vector calculus second derivative and is a scalar. The function of which the Laplacian is a second derivative is also a scalar, but the first derivative in between is the gradient, a vector. By definition,

$$\text{The Laplacian of } f(x, y, z) \equiv \text{div [grad } (f)]$$

$$\equiv \nabla \cdot [\nabla f(x, y, z)] \equiv \nabla^2 f(x, y, z) = \frac{\partial^2 f}{\partial x^2} + \frac{\partial^2 f}{\partial y^2} + \frac{\partial^2 f}{\partial z^2}$$

This derivative is useful not only in the study of electromagnetism but also in the study of the propagation of all kinds of waves.

All four of these vector calculus derivatives are very useful in the general theory of electric and magnetic fields. For example: (1) the electrostatic field $\mathbf{E}$ is related to the gradient of the electrostatic potential V: $\mathbf{E} = -\nabla V$. (2) Gauss' law is very concisely stated in terms of the divergence of the electric field, $\nabla \cdot \mathbf{E}$. (3) One of the most useful and most fundamental facts about an electrostatic field is that its curl ($\nabla \times \mathbf{E}$) is always zero everywhere. That means, for example, that if a particular electric field **E** has a nonzero curl somewhere, then the charges producing **E** are changing or moving. When **E** has a nonzero curl, **E** is (at least partly) due to magnetic effects (or represents an electromagnetic wave). (4) The general electrostatic problem can be stated in terms of the Laplacian. If you wish to find the electrostatic potential V throughout a charge-free region in space, you solve Laplace's equation $\nabla^2 V(x,y,z) = 0$ with boundary conditions at the edges of the region. If there is some (volume) charge density $\rho(x,y,z)$ in the region, you solve Poisson's equation, $\nabla^2 V(x,y,z) = -\rho(x,y,z)/\epsilon_0$.

5.2 Basic Equations

The following basic equations are discussed in intermediate-level textbooks.

$$\nabla \cdot \mathbf{E}(x,y,z) = -\frac{\rho(x,y,z)}{\epsilon_0} = 0 \quad \text{for charge-free regions} \qquad \textit{(Gauss' law)}$$

For electrostatic fields:

$$\mathbf{E}(x,y,z) = -\nabla V(x,y,z)$$

$$\nabla \times \mathbf{E}(x,y,z) = 0$$

$$\nabla^2 V(x,y,z) = -\frac{\rho(x,y,z)}{\epsilon_0} = 0 \quad \text{for charge-free regions}$$

There are a number of similar relationships in magnetism and in the relationships between electric and magnetic fields, but we will concentrate on the equations listed to demonstrate the role of the computer.

5.3 Practice in Vector Calculus

The first thing the computer allows you to do is to practice taking the vector calculus derivatives. You can calculate the values of the derivatives at a large number of points and get a feeling for the meanings of the relationships. For the greatest generality, you can take the (partial) derivatives analytically and then work with closed-form equations for the vector calculus derivatives.

Program 5.1(a)

```
GRADBAS    Gradient of function F(r)

100 DEF FNF(X,Y,Z)=1/SQR(X*X+Y*Y+Z*Z)    F(r)
110 LET D=2↑(-20)    Interval size
120 PRINT "FIELD POINT";
130 INPUT X0,Y0,Z0
140 LET G1=(FNF(X0+D,Y0,Z0)-FNF(X0-D,Y0,Z0))/(2*D)  ⎫ Gradient
150 LET G2= (FNF(X0,Y0+D,Z0)-FNF(X0,Y0-D,Z0))/(2*D) ⎬ (central
160 LET G3=(FNF(X0,Y0,Z0+D)-FNF(X0,Y0,Z0-D))/(2*D)  ⎭ difference)
170 PRINT "GRAD=";G1;G2;G3
180 PRINT
190 GOTO 120
200 END
```

Program 5.1(b)

```
GRADFOR    Gradient of function F(r)

           F(X,Y,Z)=1/SQRT(X*X+Y*Y+Z*Z)    F(r)
           D=2.**(-20)    Interval size
10         READ (5,100) X0,Y0,Z0    Field point
           IF (X0-999.) 20,90,20
20         GX=(F(X0+D,Y0,Z0)-F(X0-D,Y0,Z0))/(2.*D)  ⎫ Gradient
           GY=(F(X0,Y0+D,Z0)-F(X0,Y0-D,Z0))/(2.*D)  ⎬ (central
           GZ=(F(X0,Y0,Z0+D)-F(X0,Y0,Z0-D))/(2.*D)  ⎭ difference)
           PRINT 101,X0,Y0,Z0
           PRINT 102,GX,GY,GZ
           GOTO 10
90         STOP
100        FORMAT(3F10.5)
101        FORMAT(1X,3F10.4)
102        FORMAT(1X,5HGRAD=,3E12.4)
           END
```

The program GRAD [Programs 5.1 (a), (b)] illustrates one way to calculate the values of (the components of) the gradient of a scalar function of three variables. The function shown in the program is the electrostatic potential due to a positive point charge (of magnitude $q = \epsilon_0$) at the origin. Since you know that the electric field **E** points radially outward, the gradient at any point in this case must point toward the origin.

To take derivatives numerically, you approximate the derivative by a difference quotient. For example,

$$\frac{\partial f}{\partial x} \approx \frac{[f(x + \Delta x, y, z) - f(x - \Delta x, y, z)]}{2\,\Delta x}$$

By definition, the partial derivative is the limit of the right side as Δx goes to zero. This particular difference quotient uses central differences (f evaluated ahead and behind the point in question). Central difference approximations are practically always better than forward difference or backward difference quotients because central differences give closer values to the derivative for larger Δx's. The partial derivatives with respect to y and z are approximated in the same way. You can calculate values of the components of the gradient numerically even if you can not write an equation for the

Program 5.2(a)

```
DIVCURBA    Divergence and curl of vector field, F(r)

100 DEF FNX(X,Y,Z)=X/(X*X+Y*Y+Z*Z)↑(3/2)      Fx  }
110 DEF FNY(X,Y,Z)=Y/(X*X+Y*Y+Z*Z)↑(3/2)      Fy  } F(r)
120 DEF FNZ(X,Y,Z)=Z/(X*X+Y*Y+Z*Z)↑(3/2)      Fz  }
130 LET D=1/1024      Space increment for derivative
140 PRINT "FIELD POINT (X,Y,Z)";
150 INPUT X0,Y0,Z0
160 LET D0=(FNX(X0+D,Y0,Z0)-FNX(X0-D,Y0,Z0))/(2*D)        } Divergence
170 LET D0=D0+(FNY(X0,Y0+D,Z0)-FNY(X0,Y0-D,Z0))/(2*D)     } (central
180 LET D0=D0+(FNZ(X0,Y0,Z0+D)-FNZ(X0,Y0,Z0-D))/(2*D)     } difference)
190 PRINT "DIV =";D0
200 LET C1=(FNZ(X0,Y0+D,Z0)-FNZ(X0,Y0-D,Z0))/(2*D)        }
210 LET C1=C1-(FNY(X0,Y0,Z0+D)-FNY(X0,Y0,Z0-D))/(2*D)     }
220 LET C2=(FNX(X0,Y0,Z0+D)-FNX(X0,Y0,Z0-D))/(2*D)        } Curl
230 LET C2=C2-(FNZ(X0+D,Y0,Z0)-FNZ(X0-D,Y0,Z0))/(2*D)     } (central
240 LET C3=(FNY(X0+D,Y0,Z0)-FNY(X0-D,Y0,Z0))/(2*D)        } difference)
250 LET C3=C3-(FNX(X0,Y0+D,Z0)-FNX(X0,Y0-D,Z0))/(2*D)     }
260 PRINT "CURL =";C1;C2;C3
270 PRINT
280 GOTO 140      Return for new field point, r
290 END
```

Program 5.2(b)

```
DIVCURFO    Divergence and curl of vector field, F(r)

          FX(X,Y,Z)=X/(X*X+Y*Y+Z*Z)**1.5      Fx  }
          FY(X,Y,Z)=Y/(X*X+Y*Y+Z*Z)**1.5      Fy  } F(r)
          FZ(X,Y,Z)=Z/(X*X+Y*Y+Z*Z)**1.5      Fz  }
          D=1./1024.     Space increment
 10       READ(5,100)X0,Y0,Z0      Field point, r
          IF (X0-999.) 20,90,20
 20       D0=(FX(X0+D,Y0,Z0)-FX(X0 -D,Y0,Z0))/(2.*D)         } Divergence
          D0=D0+(FY(X0,Y0+D,Z0)-FY(X0,Y0-D,Z0))/(2.*D)      } (central
          D0=D0+(FZ(X0,Y0,Z0+D)-FZ(X0,Y0,Z0-D))/(2.*D)      } difference)
          CX=(FZ(X0,Y0+D,Z0)-FZ(X0,Y0-D,Z0))/(2.*D)         }
          CX=CX-(FY(X0,Y0,Z0+D)-FY(X0,Y0,Z0-D))/(2.*D)      }
          CY=(FX(X0,Y0,Z0+D)-FX(X0,Y0,Z0-D))/(2.*D)         } Curl
          CY=CY-(FZ(X0+D,Y0,Z0)-FZ(X0-D,Y0,Z0))/(2.*D)      } (central
          CZ=(FY(X0+D,Y0,Z0)-FY(X0-D,Y0,Z0))/(2.*D)         } difference)
          CZ=CZ-(FX(X0,Y0+D,Z0)-FX(X0,Y0-D,Z0))/(2.*D)      }
          PRINT 101,X0,Y0,Z0
          PRINT 102,D0,CX,CY,CZ
          GOTO 10     Return for new field point
 90       STOP
 100      FORMAT(3F10.5)
 101      FORMAT(1X,3F10.4)
 102      FORMAT(1X,6H  DIV=,E10.4,7H  CURL=,3E10.4)
          END
```

scalar function. The scalar function can be simply a set of numbers—as it would be, for example, if you were finding the electric field from a set of measured electrostatic potentials.

The program DIVCUR [Programs 5.2 (a), (b)] shows one way to find divergences and curls of vector fields. The x, y, and z components of the vector field **F** are functions in the program. The vector field chosen as an illustration in the program is the electric field due to a positive point charge at the origin. Both the divergence and the curl should be zero everywhere (except at the origin itself).

5.4 Solving General Electrostatic Problems

The general electrostatic problem can be stated in terms of Poisson's equation, $\nabla^2 V = -\rho/\epsilon_0$. This equation is simply a combination of Gauss' law and the fact that $\mathbf{E} = -\nabla V$. For charge-free regions, the equation reduces to Laplace's equation, $\nabla^2 V = 0$. In the most common situation, you know the potential everywhere around the surface of the region you are interested in; you use Poisson's or Laplace's equation to find V everywhere inside the region. In general, the solution to partial differential equations is complicated whether you find the solution numerically or analytically. It happens that Laplace's equation (and Poisson's equation, too, for many situations) is relatively simple and an adequate illustration of one way to treat partial differential equations on the computer. Again you approximate partial derivatives by difference quotients. Table 5.1 shows two kinds of central difference approximations, the simplest one and one which gives even better values but needs more information for each calculation.

Table 5.1 Difference Approximations

Simplest central difference approximations:

$$\frac{\partial f}{\partial x} \approx \frac{f(x+\Delta x, y, z) - f(x-\Delta x, y, z)}{2\Delta x}$$

$$\frac{\partial^2 f}{\partial x^2} \approx \frac{f(x+\Delta x, y, z) - 2f(x, y, z) + f(x-\Delta x, y, z)}{(\Delta x)^2}$$

Higher-order central difference approximations:

$$\frac{\partial f}{\partial x} \approx \frac{-f(x+2\Delta x, y, z) + 8f(x+\Delta x, y, z) - 8f(x-\Delta x, y, z) + f(x-2\Delta x, y, z)}{(12\Delta x)}$$

$$\frac{\partial^2 f}{\partial x^2} \approx \frac{-f(x+2\Delta x, y, z) + 16f(x+\Delta x, y, z) - 30f(x, y, z) + 16f(x-\Delta x, y, z) - f(x-2\Delta x, y, z)}{12(\Delta x)}$$

Using the difference equation you can find an expression for the value at one point in terms of values nearby. For simplicity, consider a two-dimensional region. Break the region up into a grid of points with equal grid spacing Δx and Δy. The value of V at a point (x, y) can be given in terms of nearby grid-point values. For example, using the simplest central difference approximation, Laplace's equation becomes (with $\Delta x = \Delta y = D$)

$$\frac{V(x+D, y) + V(x, y+D) + V(x-D, y) + V(x, y-D) - 4V(x, y)}{D^2} \approx 0$$

or

$$V(x,y) \approx \frac{V(x+D,y) + V(x,y+D) + V(x-D,y) + V(x,y-D)}{4}$$

This last equation says that the value of V at a point on the grid is approximately the average of the values at the four nearest neighbors.

The program LAPLAC [Programs 5.3 (a), (b)] illustrates one way to solve Laplace's equation in this manner. The potential on the edge of the rectangular region varies linearly. Initially, all the values of V inside the region are set to the average value on the boundary. The difference equation for V at each grid point is used over and over again at all interior points of the grid until the difference between iterations is small enough. The boundary points are always at the initially defined values.

Program 5.3(a)

```
LAPLACBA    2D Laplace's equation solution

100 DIM V(10,10),U(10,10)
110 LET I0=10      Number of intervals in x
120 LET J0=10      Number of intervals in y
130 FOR I=1 TO I0                    ⎫
140 LET V(I,1)=I                     ⎪
150 LET V(I,J0)=J0+(I-1)             ⎪
160 LET V0=V0+V(I,1)+V(I,J0)         ⎪
170 NEXT I                           ⎬ Set boundary
180 FOR J=2 TO J0-1                  ⎪ potentials
190 LET V(1,J)=J                     ⎪
200 LET V(I0,J)=I0+(J-1)             ⎪
210 LET V0=V0+V(1,J)+V(I0,J)         ⎪
220 NEXT J                           ⎭
230 LET V0=V0/(2*(I0+J0))     Average of V on boundary
240 FOR I=2 TO I0-1 ⎫ Initialize interior
250 FOR J=2 TO J0-1 ⎪ potentials to
260 LET V(I,J)=V0   ⎬ average on
270 NEXT J          ⎪ boundary
280 NEXT I          ⎭
290 LET V1=0
300 FOR I=1 TO I0       ⎫ Save values
310 FOR J=1 TO J0       ⎪ of potential
320 LET U(I,J)=V(I,J)   ⎬ from last
330 NEXT J              ⎪ iteration
340 NEXT I              ⎭
350 FOR I=2 TO I0-1 ⎫
360 FOR J=2 TO J0-1 ⎭ Step through interior points
370 LET V(I,J)=(U(I+1,J)+U(I-1,J)+U(I,J+1)+U(I,J-1))/4     New potential
380 IF V1>ABS((V(I,J)-U(I,J))/(V(I,J)) THEN 400  ⎫ Save largest
390 LET V1=ABS((V(I,J)-U(I,J))/V(I,J))           ⎭ fractional change
400 NEXT J
410 NEXT I
420 LET N=N+1      Number of iterations so far
430 IF V1>.001 THEN 290     Test for 0.1% accuracy
440 PRINT "# OF ITERATIONS OVER GRID =";N  ⎫
450 FOR J=J0 TO 1 STEP -1                  ⎪
460 FOR I=1 TO I0                          ⎪
470 PRINT V(I,J),                          ⎬ Print
480 NEXT I                                 ⎪ group
490 PRINT                                  ⎪
500 NEXT J                                 ⎭
510 END
```

Program 5.3(b)

```
LAPLACFO    2D Laplace's equation solution

         DIMENSION V(10,10),U(10,10)
         I0=10     Number of intervals in x
         J0=10     Number of intervals in y
         VAV=0.
         DO 10 I=1,I0                       }
         V(I,1)=I                           }
         V(I,J0)=J0+I-1                     }
10       VAV=VAV+V(I,1)+V(I,J0)             } Set
         DO 20 J=2,J0-1                     } boundary
         V(1,J)=J                           } potentials
         V(I0,J)=I0+J-1                     }
20       VAV=VAV+V(1,J)+V(I0,J)             }
         VAV=VAV/(2.*(I0+J0))     Average of V on boundary
         DO 30 I=2,I0-1    } Initialize interior
         DO 30 J=2,J0-1    } potentials to average
30       V(I,J)=VAV        } on boundary
         NUMIT=0
35       DVMAX=0.
         DO 40 I=1,I0      } Save values
         DO 40 J=1,J0      } of potential
40       U(I,J)=V(I,J)     } from last iteration
         DO 50 I=2,I0-1    } Step through interior values
         DO 50 J=2,J0-1    }
         V(I,J)=(U(I+1,J)+U(I-1,J)+U(I,J+1)+U(I,J-1))/4.        New potential
50       DVMAX=AMAX1(DVMAX,ABS((V(I,J)-U(I,J))/(V(I,J)))       Save largest
         NUMIT=NUMIT+1     Number of iterations so far            fractional change
         IF (DVMAX-.001) 55,55,35      Test for 0.1% accuracy
55       PRINT 100,NUMIT               }
         DO 60 J=1,J0                  } Print
         DO 60 I=1,I0                  } group
60       PRINT 101,I,J,V(I,J)          }
         STOP
100      FORMAT(1X,27H# OF ITERATIONS OVER GRID =,I3)
101      FORMAT(1X,2I4,F10.4)
         END
```

A similar technique can be used to solve Poisson's equation. You use the same ideas to approximate $\nabla^2 V$ but add at each point the amount of charge density there. The difference equation becomes

$$V(x,y) \approx \frac{V(x+D,y) + V(x,y+D) + V(x-D,y) + V(x,y-D) + (D^2/\epsilon_0)\rho(x,y)}{4}$$

The program POISSO [Programs 5.4 (a), (b)] illustrates the technique in a rectangular region with its edges held at $V \equiv 0$. ρ/ϵ_0 is chosen to be 1 everywhere.

The solution to Laplace's and Poisson's equations in multiply connected regions (regions with holes) is more complicated. In some cases the problem can be solved. If you are interested in pursuing the subject further, you can look at B. Carnahan, H. A. Luther, and J. O. Wilkes, *Applied Numerical Methods,* John Wiley & Sons, New York, 1969. Such multiply connected region problems are common. Suppose that you want to put a point charge in an otherwise charge-free region which has fixed

Program 5.4(a)

```
POISSOBA     Poisson's equation solution

100 DIM V(10,10),U(10,10)
110 LET I0=10        Number of intervals in x
120 LET J0=10        Number of intervals in y
130 FOR I=1 TO I0                      }
140 LET V(I,1)=I                       }
150 LET V(I,J0)=J0+(I-1)               } Set
160 LET V0=V0+V(I,1)+V(I,J0)           } values
170 NEXT I                             } of
180 FOR J=2 TO J0-1                    } potential V
190 LET V(1,J)=J                       } on
200 LET V(I0,J)=I0+(J-1)               } boundary
210 LET V0=V0+V(1,J)+V(I0,J)           } of grid
220 NEXT J                             }
230 LET V0=V0/(2*(I0+J0))      Average of V on boundary
240 FOR I=2 TO I0-1    } Initialize
250 FOR J=2 TO J0-1    } interior points
260 LET V(I,J)=V0      } to average on
270 NEXT J             } boundary
280 NEXT I             }
290 LET V1=0
300 FOR I=1 TO I0          }
310 FOR J=1 TO J0          } Save values
320 LET U(I,J)=V(I,J)      } from last
330 NEXT J                 } iterations
340 NEXT I                 }
350 FOR I=2 TO I0-1    } Step through interior points
360 FOR J=2 TO J0-1    }
370 LET R=1       ρ(x, y)/ε0
380 LET V(I,J)=(U(I+1,J)+U(I-1,J)+U(I,J+1)+U(I,J-1)+R)/4      New V
390 IF V1 >ABS((V(I,J)-U(I,J))/V(I,J)) THEN 410   } Find maximum
400 LET V1=ABS((V(I,J)-U(I,J))/V(I,J))            } change
410 NEXT J
420 NEXT I
430 LET N=N+1      Count iterations
440 IF V1 >.001 THEN 290      Test for 0.1% accuracy
450 PRINT "# OF ITERATIONS OVER GRID =";N   }
460 FOR J=J0 TO 1 STEP -1                   }
470 FOR I=1 TO I0                           }
480 PRINT V(I,J),                           } Print
490 NEXT I                                  } group
500 PRINT                                   }
510 NEXT J                                  }
520 END
```

Program 5.4(b)

```
POISSOFO    Poisson's equation solution

        DIMENSION V(10,10),U(10,10)
        I0=10    Number of intervals in x
        J0=10    Number of intervals in y
        VAV=0.
        DO 10 I=1,I0                   }
        V(I,1)=I                       }  Set
        V(I,J0)=J0+I-1                 }  values of
10      VAV=VAV+V(I,1)+V(I,J0)         }  potential V
        DO 20 J=2,J0-1                 }  on
        V(1,J)=J                       }  boundary of
        V(I0,J)=I0+J-1                 }  grid
20      VAV=VAV+V(1,J)+V(I0,J)         }
        VAV=VAV/(2.*(I0+J0))     Average V on boundary
        DO 30 I=2,I0-1   } Initialize interior
        DO 30 J=2,J0-1   } points to average
30      V(I,J)=VAV       } V on boundary
        NUMIT=0
35      DVMAX=0.
        DO 40 I=1,I0     } Save
        DO 40 J=1,J0     } old
40      U(I,J)=V(I,J)    } values
        DO 50 I=2,I0-1   }
        DO 50 J=2,J0-1   } Step through interior points
        RHO=.5     ρ(x, y)/ε0
        V(I,J)=(U(I+1,J)+U(I-1,J)+U(I,J+1)+U(I,J-1)+RHO)/4.    New V
50      DVMAX=AMAX1(DVMAX,ABS((V(I,J)-U(I,J))/V(I,J)))    Find maximum change
        NUMIT=NUMIT+1    Count iterations
        IF (DVMAX-.001) 55,55,35    Test for 0.1% accuracy
55      PRINT 100,NUMIT          }
        DO 60 J=1,J0             }  Print
        DO 60 I=1,I0             }  group
60      PRINT 101,I,J,V(I,J)     }
        STOP
100     FORMAT(1X,27H# OF ITERATIONS OVER GRID =,I3)
101     FORMAT(1X,2I4,F10.4)
        END
```

potentials on its boundary. One way to solve the problem is to fix the potential on the grid points near the point charge as being due entirely to the point charge. You then have to solve Laplace's equation in a region having a hole in it—the hole is the region around the point charge. You must still assume that all the values of V on the boundary are fixed, even those values right around the hole.

Problems

The following represent ways the material in this chapter has been used.

1. Consider a two-dimensional charge-free square region of space with linearly increasing potential on its edges.

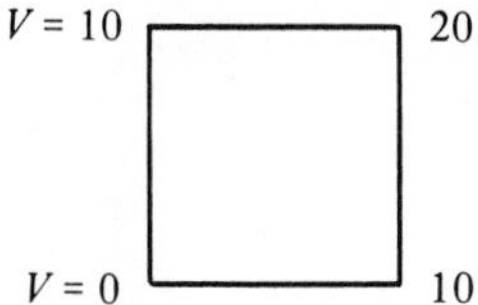

   a. Using a program which solves Laplace's equation, find the interior potentials to 0.1% accuracy.
   b. Vary the number of grid points covering the region and find out how quickly (in terms of number of iterations over the grid and computer time used) you reach 0.1% accuracy.
   c. For one number of grid points, vary the accuracy you demand and investigate how long the calculation takes.

2. Consider a two-dimensional charge-free rectangular region of space with constant potentials on the sides.

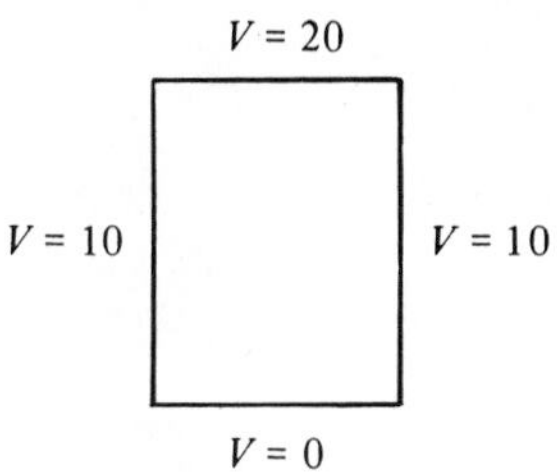

   a. Using a program which solves Laplace's equation, find the interior potentials to 0.1% accuracy.
   b. Vary the number of grid points covering the region and find out how quickly (in terms of number of iterations over the grid and computer time used) you reach 0.1% accuracy.
   c. For one number of grid points, vary the accuracy you demand and investigate how long the calculation takes.

3. Consider a two-dimensional square region of space with a uniform charge density $\rho = \epsilon_0$ throughout the region and with the edges of the region held at $V = 0$.
   a. Using a program which solves Poisson's equation, find the interior potentials to 0.1% accuracy.
   b. Now allow the charge density to vary with position as $\rho(x,y) = \epsilon_0 xy$. Find the interior potentials to 0.1%.

4. As pointed out in the chapter, higher-convergence approximations to the second partial derivatives in Laplace's equation are possible.
   a. Write a program using a better approximation. (You will need to use the simpler approximation for interior points near the boundary.)
   b. Solve for the interior potentials (to 0.1%) of a charge-free square region with a constant potential on the boundary. Compare your results with the better approximation to those using the simpler approximation. Consider the number of iterations for a given accuracy and the computer time used.
   c. Generalize your program to solve Poisson's equation.

6 magnetic fields

6.1 Introduction

Static magnetic fields are caused by steady currents through wires. The wires may be straight or bent; coils of wire are commonly used to produce magnetic fields. As in electrostatics, the basic equations of magnetostatistics are fairly complicated partial differential equations. Most visualization of physical situations is done in terms of magnetic field lines. (A magnetic potential is also associated with magnetic fields, but it turns out to be a vector field itself.) If the currents vary sufficiently slowly, even time-varying magnetic fields can be calculated in the static-field manner.

Magnetic field lines (flux lines, to be more precise) are defined just as are electrostatic field lines. The only difference is that **B** (or **H**) is mapped instead of **E**. Field lines are associated in the same way with almost any vector field. The field lines are defined by the following two properties:

I. At any point, the tangent to the field line is parallel to the field at that point.

II. At any point, the number of field lines (crossing a unit cross section placed perpendicular to the lines) is proportional to the magnitude of the field there.

Property I provides a simple way to follow magnetic field lines. A complete discussion of the method of calculating field lines is given in Chapter 4. Property I demands that (in two dimensions, for simplicity) the triangle formed by Δx, Δy, and Δs be similar to the triangle formed by B_x, B_y and $\sqrt{B_x^2 + B_y^2}$. If you walk along a field line in steps Δs, then the components of Δs parallel to the axes are given by

$$\Delta x = \Delta s \frac{B_x}{\sqrt{B_x^2 + B_y^2}} \qquad \Delta y = \Delta s \frac{B_y}{\sqrt{B_x^2 + B_y^2}}$$

One difference between electrostatics and magnetostatics is that you cannot separate a positive magnetic charge from a negative magnetic charge. Hence, unlike electric field lines, which can end on charges, magnetic field lines never end at all.

Coils and wires are emphasized in this chapter. The magnetism from pieces of iron will not be discussed. [The solid-state physicist maintains that the ferromagnetism of materials like iron can be described as due to very small (microscopic) looplike currents.] The subject will be developed in terms of the magnetic field of a wire, then a loop, and then combinations of loops.

6.2 Basic Equations

The following basic relationships are discussed in your textbook. If you do not recognize the laws, review your text.

1. $\mathbf{F} = q_0(\mathbf{E} + \mathbf{V} \times \mathbf{B})$ *(Definition of* **B***)*

   F = the total force on test charge q_0 moving at velocity **V**. The total force on a point charge can always be written this way. **E** is called the electric field and **B** is called the magnetic field.

2. $\mathbf{B} = \dfrac{\mu_0 I}{2\pi r}$ *(Tangential; sense by the right-hand rule)*

   B = the field due to a long straight wire carrying current I.

3. $\oint \mathbf{B} \cdot d\mathbf{l} = \mu_0 I$ *(Ampere's law)*

   $\oint$ means an integral around a closed path; I is the net enclosed current.

4. $d\mathbf{B} = \dfrac{\mu_0 I}{4\pi} \dfrac{d\mathbf{l} \times \mathbf{r}}{r^3}$ *(Biot-Savart law)*

   where $d\mathbf{B}$ = magnetic field at point **r** due to the small wire segment **dl** carrying current I
   r = vector from **dl** to the observation point
   $\mathbf{B} = \int d\mathbf{B}$ determines the total field due to all the wire segments

5. $\mathcal{E} = \int \mathbf{E} \cdot d\mathbf{l} = -\dfrac{d\phi}{dt}$ *(Faraday's law)*

$\mathcal{E}$ is the emf developed around a closed contour due to changes in the magnetic flux ϕ linking the contour.

6.3 Long Straight Wires

The field due to sets of long straight wires is easy to calculate. A program which follows field lines due to such sets of long straight wires is BWIRE [Programs 6.1(a), (b)].

6.4 Loop-Field Calculations

This section has to do with calculating the magnetic field at any point in space due to a single circular loop of wire and then calculating more complicated fields by putting

Program 6.1(a)

```
BWIREBAS   Flux lines for sets of ∞ wires parallel to Z

100 LET S0=.02     Step size Δs
110 PRINT "# OF WIRES (PERP. TO PAGE)";   }
120 INPUT N0                              }
130 FOR J=1 TO N0                         } Geometry
140 PRINT "X,Y,I FOR WIRE";J;             } of ∞
150 INPUT X(J),Y(J),I(J)                  } wires
160 NEXT J                                }
170 PRINT "STARTING POINT X,Y";
180 INPUT X0,Y0
190 PRINT "X","Y"
200 LET X1=X0  }
210 LET Y1=Y0  }
220 LET X2=0   } Initialization
230 LET Y2=0   }
240 LET B1=0   }
250 LET B2=0   }
260 FOR J=1 TO N0      Sum over wires
270 LET X9=Y1+Y2/2-Y(J)        } r from wire
280 LET Y9=-(X1+X2/2-X(J))     } to field point
290 LET R2=X9*X9+Y9*Y9         } (half-stepped)
300 LET B1=B1+2E-7*I(J)*X9/R2    Bx
310 LET B2=B2+2E-7*I(J)*Y9/R2    By
320 NEXT J
330 LET B0=SQR(B1*B1+B2*B2)   |B|
340 LET X2=S0*B1/B0   Δx
350 LET Y2=S0*B2/B0   Δy
360 LET X1=X1+X2     New x
370 LET Y1=Y1+Y2     New y
380 PRINT X1,Y1
390 IF ABS(X1-X0)+ABS(Y1-Y0)<.9*S0 THEN 170   Back to start?  } Tests for
400 IF ABS(X1)+ABS(Y1)>2 THEN 170   Far off page?             } end of
410 GOTO 240     Return for new flux line                      } flux line
420 END
```

Program 6.1(b)

```
BWIREFOR    Flux lines for sets of ∞ wires parallel to Z

                                                         Geometry of wires
        DIMENSION X(10),Y(10),CUR(10)
        DATA NWIR,X(1),Y(1),CUR(1),X(2),Y(2),CUR(2)/2,1.,0.,1.,-1.,0.,1./
        DS=.02      Step size, Δs
10      READ(5,100)XST,YST      Starting point
        IF (XST-999.) 15,90,15
15      X1=XST  ┐
        Y1=YST  │
        DX=0.   │
        DY=0.   ├ Initialization
20      BX=0.   │
        BY=0.   ┘
        DO 30 J=1,NWIR      Sum over wires
        X9=Y1+DY/2.-Y(J)        } r from each wire
        Y9=-(X1+DX/2.-X(J))     } to field point (half-stepped)
        R2=X9*X9+Y9*Y9
        BX=BX+2.E-7*CUR(J)*X9/R2    Bx  ┐
30      BY=BY+2.E-7*CUR(J)*Y9/R2    By  ├ B
        B=SQRT(BX*BX+BY*BY)     |B|     ┘
        DX=DS*BX/B      Δx
        DY=DS*BY/B      Δy
        X1=X1+DX        New x
        Y1=Y1+DY        New y
        PRINT 101,X1,Y1
        IF (ABS(X1-XST)+ABS(Y1-YST)-.9*DS) 10,20,20 Back to start? Then start new line
90      STOP
100     FORMAT(2F10.5)
101     FORMAT(1X,2F10.4)
        END
```

several loops together to form coils. You can do the same sort of thing by starting with the calculations of **B** due to a short straight wire segment carrying current. Short straight wire segments go together to form square coils.

The calculation of the magnetic field at any point in space is based on the Biot-Savart law. Although the Biot-Savart law is practically always mentioned in introductory physics, it is difficult to use in hand calculations. For computer calculations, it is a reasonable starting point.

COIL [Programs 6.2(a), (b)] and BLINE [Programs 6.3(a), (b), (c)] are programs which have been used to calculate magnetic fields. The fundamental routine uses the following method. Consider a circular loop of wire [radius A; parallel to xy plane;

Program 6.2(a)

```
COILBAS   B fields for sets of coils parallel to YZ plane

100 LET N0=2      Number of coils
110 FOR J=1 TO N0                    }
120 READ X(J),R(J),I(J)              } Geometry of
130 NEXT J                           } coils
140 DATA .5,1,1, -.5,1,1             }
150 LET P2=6.28318530     2π
160 LET N9=10     Number of segments for each coil
170 FOR I=1 TO N9                    }
180 LET T=P2*(I-.5)/N9               } Store sines
190 LET C(I)=COS(T)                  } and cosines
200 LET S(I)=SIN(T)                  } used later
210 NEXT I                           }
220 PRINT "(X,Y,Z) FIELD POINT?"
230 INPUT X0,Y0,Z0
240 LET B1=0   }
250 LET B2=0   } Initialization
260 LET B3=0   }
270 FOR J=1 TO N0      Sum over coils
280 FOR I=1 TO N9      Sum over segments of each coil
290 LET L1=0                 }
300 LET L2=-P2*R(J)*S(I)/N9  } dl
310 LET L3=P2*R(J)*C(I)/N9   }
320 LET L7=X(J)       }
330 LET L8=R(J)*C(I)  } Coordinates
340 LET L9=R(J)*S(I)  } of dl
350 LET X6=X0-L7  }
360 LET Y6=Y0-L8  } r from segment
370 LET Z6=Z0-L9  } to field point
380 LET R6=SQR(X6*X6+Y6*Y6+Z6*Z6)
390 LET R6=R6*R6*R6     |r|^3
400 LET C1=L2*Z6-L3*Y6  }
410 LET C2=L3*X6-L1*Z6  } dl x r
420 LET C3=L1*Y6-L2*X6  }
430 LET B1=B1+1E-7*I(J)*C1/R6    Bx
440 LET B2=B2+1E-7*I(J)*C2/R6    By
450 LET B3=B3+1E-7*I(J)*C3/R6    Bz
460 NEXT I
470 NEXT J
480 PRINT B1;B2;B3
490 PRINT
500 GOTO 220      Return for next field point
510 END
```

Program 6.2(b)

```
COILFOR    B fields for sets of coils parallel to XY plane

           DATA CURENT,RADIUS,NCOIL,SEP/1.,.01,2,.005/      Geometry of coils
   10      READ (5,101)X,Y,Z     Field point
           BX=0.
           BY=0.
           BZ=0.
           DO 20 N=1,NCOIL,1     Sum over coils
           AN=N
           ZCOIL=(AN-1.)*SEP     Z of coil
           BXCOIL=0.
           BYCOIL=0.
           BZCOIL=0.
           DO 30 I=1,16,1     Sum over segments of each coil
           AI=I
           ANG=(AI-1.)*3.14159/16.
           COSANG=COS(ANG)
           SINANG=SIN(ANG)
           DLX=-6.28318*RADIUS*SINANG/32.  }
           DLY=6.28318*RADIUS*COSANG/32.   } dl
           DLZ=0.                          }
           XSEG=X-RADIUS*COSANG  } r from dl
           YSEG=Y-RADIUS*SINANG  } to field point
           ZSEG=Z-ZCOIL          }
           RSEG=SQRT(XSEG**2+YSEG**2+ZSEG**2)
           RCUBED=RSEG**3     |r|^3
           XCROSS=DLY*ZSEG-DLZ*YSEG  }
           YCROSS=DLZ*XSEG-DLX*ZSEG  } dl x r
           ZCROSS=DLX*YSEG-DLY*XSEG  }
           BXCOIL=BXCOIL+1.E-7*CURENT*XCROSS/RCUBED   Bx }  B for each
           BYCOIL=BYCOIL+1.E-7*CURENT*YCROSS/RCUBED   By }  coil in turn
   30      BZCOIL=BZCOIL+1.E-7*CURENT*ZCROSS/RCUBED   Bz }
           BX=BX+BXCOIL   Bx }  Total
           BY=BY+BYCOIL   By }  B field
   20      BZ=BZ+BZCOIL   Bz }
           PRINT 102,BX,BY,BZ
           IF (CURENT) 10,40,10     Return for next field point
  101      FORMAT(3F5.3)
  102      FORMAT(1X,3E15.5)
   40      STOP
           END
```

centered on the point (0, 0, z)] and some observation point (x_2, y_2, z_2) (see Figure 6.1). Break the loop up into short segments and approximate each segment by a straight line $d\mathbf{l}$. Use the Biot-Savart law to find the field due to each little segment, and then add all the fields due to all the segments. You must take the cross product in the Biot-Savart law correctly. You are using Cartesian coordinates, so the cross product is

$$(d\mathbf{l} \times \mathbf{r})_x = (d\mathbf{l})_y (\mathbf{r})_z - (d\mathbf{l})_z (\mathbf{r})_y$$

$$(d\mathbf{l} \times \mathbf{r})_y = (d\mathbf{l})_z (\mathbf{r})_x - (d\mathbf{l})_x (\mathbf{r})_z$$

$$(d\mathbf{l} \times \mathbf{r})_z = (d\mathbf{l})_x (\mathbf{r})_y - (d\mathbf{l})_y (\mathbf{r})_x$$

The result of the routine is the magnetic field (B_x, B_y, B_z) at the observation point (x_2, y_2, z_2).

You can now use this field (B_x, B_y, B_z) in a number of ways. First, you could now calculate the field line and step along the B field line to the next point. At the new point you could recall the B routine to get the field at this new point, and so on.

A second thing you could do is call the routine again for a loop at a different position. For example, adding the fields at any observation point due to two loops at (0, 0, $-A/2$) and (0, 0, $+A/2$) gives you the field due to a Helmholtz pair of radius A. You could also approximate a short solenoid by a series of closely spaced single loops (see Figures 6.2 to 6.4).

Program 6.3(a) Flow chart for a magnetic field-line strategy. After setting the initial values of parameters, the calculation walks along the field line a step at a time. The field at each point determines the components x and y of the next step.

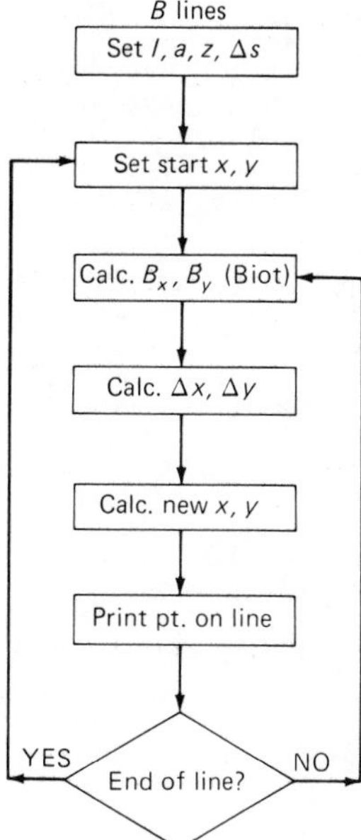

Program 6.3(b)

```
BLINEBAS    Flux lines for sets of coils parallel to YZ plane

100 LET N0=2   Number of coils
110 FOR J=1 TO N0                   }
120 READ X(J),R(J),I(J)             } Geometry of
130 NEXT J                          } coils (parallel
140 DATA .5,1,1, -.5,1,1            } to yz plane)
150 LET P2=6.28318530   2π
160 LET N9=10   Number of segments for each coil
170 FOR I=1 TO N9
180 LET T=P2*(I-.5)/N9   } Sines and
190 LET C(I)=COS(T)      } cosines needed
200 LET S(I)=SIN(T)      } in calculation
210 NEXT I               }
220 LET D=.1   Step size, Δs
230 PRINT "(X,Y) STARTING PT. ON FIELD LINE?"
240 INPUT X5,Y5
250 LET X0=X5   }
260 LET Y0=Y5   }
270 LET X1=0    }
280 LET Y1=0    }
290 LET S0=.5   } Initialization
300 LET S=0     }
310 LET B1=0    }
320 LET B2=0    }
330 LET B3=0    }
340 FOR J=1 TO N0   Sum over coils
350 FOR I=1 TO N9   Sum around each coil
360 LET L1=0                    }
370 LET L2=-P2*R(J)*S(I)/N9     } dl
380 LET L3=P2*R(J)*C(I)/N9      }
390 LET L7=X(J)        } Coordinates of
400 LET L8=R(J)*C(I)   } dl
410 LET L9=R(J)*S(I)   }
420 LET X6=X0+X1/2-L7  } r from
430 LET Y6=Y0+Y1/2-L8  } dl to field point (half-stepped)
440 LET Z6=-L9         }
450 LET R6=SQR(X6*X6+Y6*Y6+Z6*Z6)
460 LET R6=R6*R6*R6   |r|^3
470 LET C1=L2*Z6-L3*Y6  }
480 LET C2=L3*X6-L1*Z6  } dl x r
490 LET C3=L1*Y6-L2*X6  }
500 LET B1=B1+1E-7*I(J)*C1/R6   Bx }
510 LET B2=B2+1E-7*I(J)*C2/R6   By } B
520 LET B3=B3+1E-7*I(J)*C3/R6   Bz }
530 NEXT I
540 NEXT J
550 LET B0=SQR(B1*B1+B2*B2+B3*B3)   |B|
560 LET X1=D*B1/B0   dx
570 LET X0=X0+X1   New x
580 LET Y1=D*B2/B0   dy
590 LET Y0=Y0+Y1   New y
600 LET S=S+D              }
610 IF S<S0 THEN 640       } Print
620 LET S=0                } group
630 PRINT X0,Y0            }
640 IF ABS(X0)+ABS(Y0)>10 THEN 660   Far off page?             } Tests for
650 IF ABS(X0-X5)+ABS(Y0-Y5)>.9*D THEN 310   Back to start?    } end of
                                                                } flux line
660 PRINT X0,Y0   Print last point
670 GOTO 230   Return to start new line
680 END
```

Program 6.3(c)

```
BLINEFOR    Flux lines for sets of coils parallel to yz plane

            DIMENSION X(10),R(10),CUR(10),C(10),S(10)
            NCOI=2   Number of coils
            DATA X(1),R(1),CUR(1),X(2),R(2),CUR(2)/.5,1.,1.,-.5,1.,1./  Geometry of coils
            NPIEC=10   Number of segments for each coil
            TWOPI=6.2831853   2π
            DO 10 I=1,NPIEC          }
            AI=I                     } Store sines
            T=TWOPI*(AI-.5)/NPIEC    } and cosines
            C(I)=COS(T)              } used later
    10      S(I)=SIN(T)              }
            DS=.1   Step size, ΔS
    15      READ(5,100)XST,YST   Starting point
            IF (XST-999.) 20,90,90
    20      X0=XST     }
            Y0=YST     }
            DX=0.      }
            DY=0.      }
            PRNTS=.5   } Initialization
            DIST=0.    }
    25      BX=0.      }
            BY=0.      }
            BZ=0.      }
            DO 30 J=1,NCOI    Sum over coils
            DO 30 I=1,NPIEC   Sum over segments of each coil
            DLX=0.                        }
            DLY=-TWOPI*R(J)*S(I)/NPIEC    } dl
            DLZ=TWOPI*R(J)*C(I)/NPIEC     }
            XDL=X(J)           }
            YDL=R(J)*C(I)      } Coordinates of dl
            ZDL=R(J)*S(I)      }
            XX=X0+DS/2.-XDL    } r from segment
            Y=Y0+DY/2.-YDL     } to field point (half-stepped)
            Z=-ZDL             }
            RR=(XX*XX+Y*Y+Z*Z)**1.5
            CX=DLY*Z-DLZ*Y     }
            CY=DLZ*XX-DLX*Z    } dl x r
            CZ=DLX*Y-DLY*XX    }
            BX=BX+1.E-7*CUR(J)*CX/RR   Bx  }
            BY=BY+1.E-7*CUR(J)*CY/RR   By  }
    30      BZ=BZ+1.E-7*CUR(J)*CZ/RR   Bz  } B
            B=SQRT(BX*BX+BY*BY+BZ*BZ)  |B| }
            DX=DS*BX/B   dx
            X0=X0+DX     New x
            DY=DS*BY/B   dy
            Y0=Y0+DY     New y
            DIST=DIST+DS                 }
            IF (DIST-PRNTS) 50,40,40     } Print
    40      DIST=0.                      } group
            PRINT 101,X0,Y0              }
    50      IF (ABS(X0-XST)+ABS(Y0-YST)-.9*DS) 60,60,25   Back to start?  } Tests for
    60      PRINT 101,X0,Y0   Print last point                            } end of
            GOTO 15   Return for new flux line                            } flux line
    90      STOP
   100      FORMAT(2F10.5)
   101      FORMAT(1X,2F10.4)
            END
```

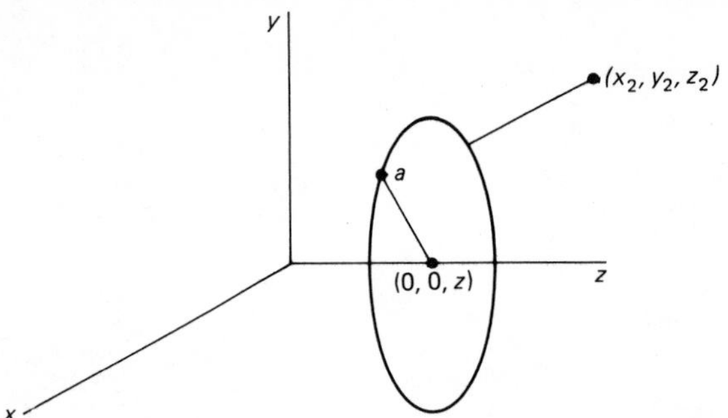

Figure 6.1 The geometry for a single circular loop parallel to the xy plane; $(0, 0, z)$ is the center of the loop; (x_2, y_2, z_2) is the point at which the field is calculated.

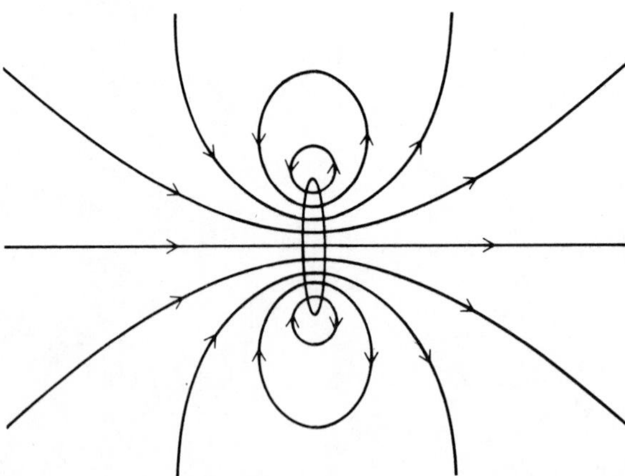

Figure 6.2 Magnetic field lines due to a single loop of wire. The loop is perpendicular to the page and is centered on the origin (the center of the page). The field lines drawn were chosen to be distributed evenly across the inside diameter of the loop. Is this the correct distribution according to Property 2 for field lines? What does the field pattern look like far from the loop? What does "far" mean in this connection?

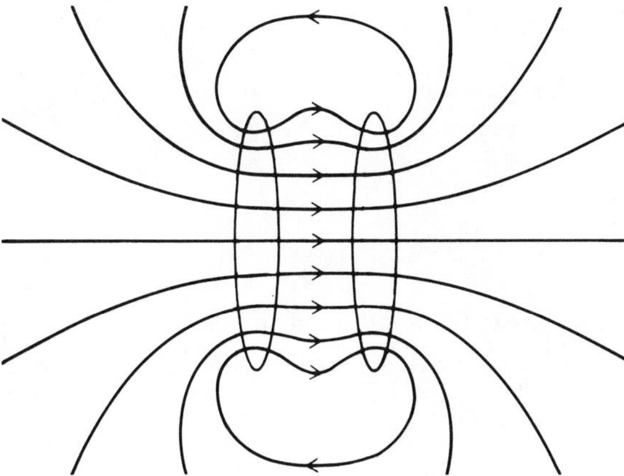

Figure 6.3 Field lines for a Helmholtz pair (two equivalent coils spaced a radius apart). Again, the coils are perpendicular to the page. The field lines were drawn so as to be equally spaced across the central plane of the Helmholtz pair. The increased region of homogeneity of the field is clearly shown.

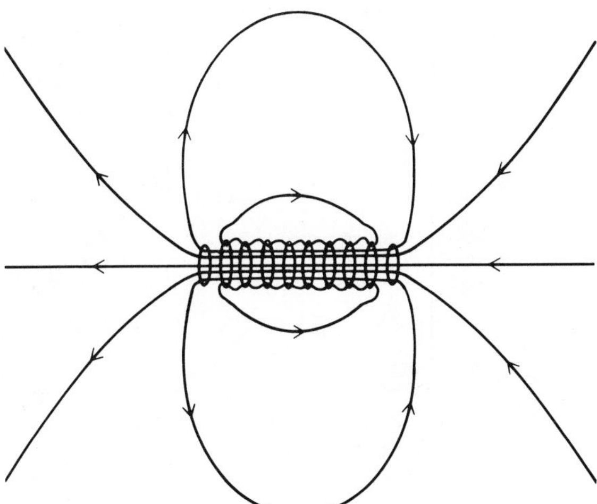

Figure 6.4 Field lines for ten single loops spaced out across the page. This is an approximation to a short solenoid. How far should you be from the loops in order for the approximation to be fairly good? Can you identify the ripple that is due entirely to the finite spacing of the loops? Since real solenoids are usually made from closely wound, insulated wires, will such a ripple occur in real solenoids? Could you design a solenoid with essentially no ripple?

6.5 Motion of Charges in Magnetic Fields

Now that you have a way to calculate magnetic fields, you can try to calculate the trajectories of charges moving in these fields. The basic program is an $\mathbf{F} = m\mathbf{a}$ program in which the force is the Lorentz force:

$$\mathbf{F} = q(\mathbf{v} \times \mathbf{B})$$

Since the force is now velocity-dependent, you must be careful in integrating $\mathbf{F} = m\mathbf{a}$. EBTRAJ [Programs 6.4(a), (b), (c)] is a program for which $\mathbf{F} = q(\mathbf{E} + \mathbf{v} \times \mathbf{B})$.

Program 6.4(a) Block diagram of the strategy which finds the path of charged particles in arbitrary combinations of electrostatic and magnetostatic fields. Since the force is velocity-dependent, the acceleration is calculated twice in each time step.

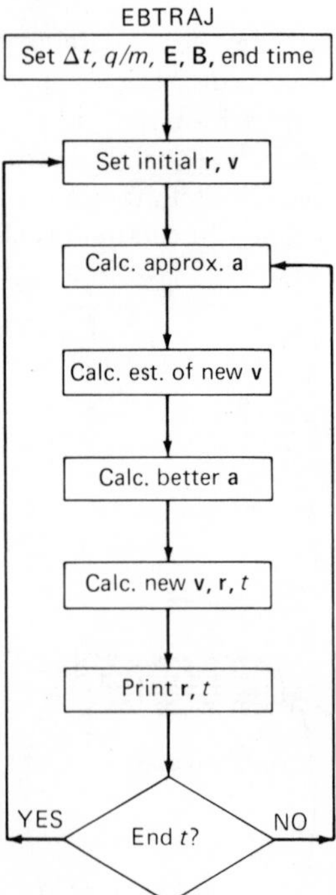

Program 6.4(b)

EBTRABA *Charged particle motion in E, B*

```
100 LET D=.01            Time step, Δt
110 LET D1=1             Print time
120 LET Q=1              q/m
130 READ E1,E2,E3,B1,B2,B3          } E and B fields
140 DATA 0,0,0,0,0,1                }
150 PRINT "INITIAL (X,Y,Z) & (VX,VY,VZ)?"
160 INPUT X1,Y1,Z1,V1,V2,V3
170 LET T=0
180 LET T0=10            End time
190 LET V0=SQR(V1*V1+V2*V2+V3*V3)
200 PRINT T,X1,Y1,Z1,V0
210 LET A1=Q*((V2*B3-V3*B2)+E1)     } First calculation
220 LET A2=Q*((V3*B1-V1*B3)+E2)     } of acceleration
230 LET A3=Q*((V1*B2-V2*B1)+E3)     }
240 IF T>0 THEN 290      }
250 LET V1=V1+A1*D/2     }
260 LET V2=V2+A2*D/2     } Initial half-step
270 LET V3=V3+A3*D/2     }
280 GOTO 380             }
290 LET U1=V1+A1*D/2     } Estimate of
300 LET U2=V2+A2*D/2     } in-step velocity
310 LET U3=V3+A3*D/2     }
320 LET A1=Q*((U2*B3-U3*B2)+E1)     } Second calculation
330 LET A2=Q*((U3*B1-U1*B3)+E2)     } of acceleration
340 LET A3=Q*((U1*B2-U2*B1)+E3)     }
350 LET V1=V1+A1*D       }
360 LET V2=V2+A2*D       } New velocity (half-stepped)
370 LET V3=V3+A3*D       }
380 LET X1=X1+V1*D       }
390 LET Y1=Y1+V2*D       } New position
400 LET Z1=Z1+V3*D       }
410 LET T=T+D            New time
420 IF ABS(T-D1)>D/2 THEN 460       }
430 LET V0=SQR(U1*U1+U2*U2+U3*U3)   } Print
440 PRINT T,X1,Y1,Z1,V0             } group
450 LET D1=D1+1                     }
460 IF T<T0 THEN 210     Return for next Δt step
470 END
```

Program 6.4(c)

```
EBTRAFO   Charged particle motion in E, B

          DT=.01     Time step, Δt
          PINT=1.    Print time
          QTOM=1.    q/m
          DATA EX,EY,EZ,BX,BY,BZ/0.,0.,0.,0.,0.,1./     E and B fields
          DATA X,Y,Z,VX,VY,VZ/-1.,0.,0.,0.,1.,0./      Initial position and velocity
          TIM=0.
          TEND=10.   End time
          VMAG=SQRT(VX*VX+VY*VY+VZ*VZ)
          PRINT 101,TIM,X,Y,Z,VMAG
5         AX=QTOM*((VY*BZ-VZ*BY)+EX)    }
          AY=QTOM*((VZ*BX-VX*BZ)+EY)    } First calculation of acceleration
          AZ=QTOM*((VX*BY-VY*BX)+EZ)    }
          IF (TIM) 10,10,20  }
10        VX=VX+AX*DT/2.     }
          VY=VY+AY*DT/2.     } Initial half-step
          VZ=VZ+AZ*DT/2.     }
          GOTO 30
20        V1X=VX+AX*DT/2.    }
          V1Y=VY+AY*DT/2.    } Estimate of in step velocity
          V1Z=VZ+AZ*DT/2.    }
          AX=QTOM*((V1Y*BZ-V1Z*BY)+EX)  }
          AY=QTOM*((V1Z*BX-V1X*BZ)+EY)  } Second calculation of acceleration
          AZ=QTOM*((V1X*BY-V1Y*BX)+EZ)  }
          VX=VX+AX*DT  }
          VY=VY+AY*DT  } New velocity (half-stepped)
          VZ=VZ+AZ*DT  }
30        X=X+VX*DT  }
          Y=Y+VY*DT  } New position
          Z=Z+VZ*DT  }
          TIM=TIM+DT       New time
          IF (ABS(TIM-PINT)-DT/2.) 40,40,50   }
40        VMAG=SQRT(V1X*V1X+V1Y*V1Y+V1Z*VIZ)  } Print group
          PRINT 101,TIM,X,Y,Z,VMAG            }
          PINT=PINT+1.                        }
50        IF (TIM-TEND) 5,60,60    Return for next Δt step
60        STOP
101       FORMAT(1X,5F10.5)
          END
```

Problems

A number of different kinds of problems are possible with this material. A few problems illustrative of the various kinds follow.

1. Calculate the total force on a unit positive point test charge under the following situations. Choose several different points in space to calculate the force.
   a. $\mathbf{E} = 0$, $\mathbf{B}$ uniform in $+Z$ direction; $v_{test} = 1$, parallel to $+x$ axis and $v_{test} = 1$, parallel to $+z$ axis.
   b. $\mathbf{E}$ due to unit positive point charge at the origin, $\mathbf{B}$ and $\mathbf{v}$ as in part a.

2. Write the force on a short straight wire segment (length $d\mathbf{l}$) carrying current I as $d\mathbf{F} = I\, d\mathbf{l} \times \mathbf{B}$

a. Calculate the total force on a rectangle of wire (sides of length 1 and 2 m) lying in the xy plane and carrying a current of 1 A. The rectangle lies in a uniform magnetic field of 1 T parallel to the $+z$ axis. Now make **B** parallel to the $+x$ axis and calculate the total force again.

b. A circular loop of wire (radius 1 m) is immersed in the uniform **B** field parallel to the $+z$ axis from part a. The loop is made to rotate around a diameter which is parallel to the x axis. Calculate the total force on the loop at a number of different times during a complete rotation. This is the force which must be used to keep a generator going.

3. Write a program to calculate the line integral of $\mathbf{B} \cdot d\mathbf{l}$ along piecewise straight paths parallel to the coordinate axes. Use this program and a magnetic field subroutine to show that Ampere's law holds for the following situations (at least along some paths of this type):
   a. The magnetic field **B** produced by a long straight wire carrying a current of 1 A.
   b. The magnetic field produced by a circular loop carrying 1 A.
   c. The magnetic field produced by some exotic current distribution of your own choosing. Can you ever prove a general result such as Ampere's law using the computer?

4. a. Write a program using the Biot-Savart law to calculate the field at any point due to a straight wire of length l parallel to the x axis.
   b. Rewrite the program as a subroutine. Allow the straight wire to lie between any points (x_1, y_1, z_1) and (x_2, y_2, z_2) which are given in the call to the subroutine.
   c. Use your subroutine to calculate the field at a number of points due to a rectangular loop of wire.
   d. Using the program from part c and a field line subroutine, calculate and plot some **B** field lines for a rectangular loop.
   e. Develop a procedure calculating the field due to a short solenoid of rectangular cross section.

5. a. Using the ideas about circular loops developed in an earlier section, write a program to calculate the magnetic field due to a quadrupole arrangement of coils.
   b. Quadrupole arrangements of magnetic fields are sometimes useful in controlling or directing beams of charged particles. Calculate the trajectories of some charged particles injected into this field configuration.

6. Faraday's law relates the emf around one circuit to the time rate of change of magnetic flux linking the circuit. Among other concepts related to Faraday's law are mutual and self inductances.
   a. Allow the current flowing in a single turn coil to vary with time as

$I = I_0 \sin \omega t$. Use a magnetic field subroutine to find the field anywhere for any time t.

b. Place a second coil of different radius and orientation in the field of the first coil. Divide the cross-sectional area of coil two into a number of pieces and calculate the flux linking the coil as

$$\phi = \sum_{i=1}^{N} \mathbf{B}_i \cdot \Delta \mathbf{A}_i$$

This flux is a function of time.

c. Find the emf in coil two due to coil one by calculating the time derivative of the flux from part b. How does the emf depend on the relative orientations of the two coils? For several positions and orientations, find the mutual inductance,

$$M = \phi / I$$

d. Now let the current in coil one be stationary, but pull coil two away from coil one at constant speed in various directions and for various relative orientations. Calculate the time rate of change of the flux through coil two and the emf developed. This is sometimes called motional emf. How does it depend on the withdrawal speed?

e. For constant current in coil one, rotate coil two, and calculate the emf. How does the emf depend on the rotation frequency? This is the basis of a motor generator.

7. Think up several problems using the computer to make some concepts clearer to you which you found especially difficult in magnetostatics. Document your solutions with a complete description of your solution procedure.

7 relativity: E and B

7.1 Introduction

It has been said that a magnetic field is an electric field seen from the wrong frame of reference. Of course, by the principle of relativity, all (inertial) frames of reference are equivalent, so there are no right or wrong frames of reference. The offhand comment merely points out that if you view a physical situation from the rest frame of a test charge, a stationary magnetic field produces no force. This is a consequence of the form of the Lorentz force $F = q(\mathbf{v}_q \times \mathbf{B})$; since the velocity $\mathbf{v}_q$ is zero in the rest frame of the test charge q, the magnetic force is zero. In the lab frame of reference, the velocity is not zero, so a magnetic force occurs. In the rest frame of the charge, the magnetic field is different from the magnetic field in the lab frame, and an electric field appears. By the principle of relativity, the physical predictions must be the same. What appears as a magnetic field in the lab frame appears as an electric field in the rest frame of the test charge. This relationship is predicted by the theory of special relativity.

You have probably already been introduced to special relativity. What this chapter does is elaborate the use of the Lorentz transformation of special relativity and discuss several interesting consequences of relativity—especially that of the relationship between **E** and **B**.

7.2 Basic Facts

The following facts are discussed in intermediate textbooks. The Lorentz transformation between any two inertial frames of reference S and S' can be put in the form

$$x' = \gamma(x - vt)$$
$$y' = y$$
$$z' = z$$
$$t' = \gamma\left(t - \frac{vx}{c^2}\right)$$

and its inverse

$$x = \gamma(x' + vt')$$
$$y = y'$$
$$z = z'$$
$$t = \gamma\left(t' + \frac{vx'}{c^2}\right)$$

where v is the velocity of S' relative to S and is positive when v points along the x axis.

$$\gamma = \frac{1}{\sqrt{1 - v^2/c^2}}$$

The x and x' (y and y'; z and z') axes have been chosen to be parallel. t and t' are set so that the times are zero when the origins coincide.

7.3 The Lorentz Contraction

The Lorentz-Fitzgerald contraction is one of the first striking consequences of special relativity. Often labeled epigrammatically as proof that "moving lengths are shorter," it is in fact much more subtle than that.

Assume a rod of length L_0 at rest in one frame, say S'. The rod lies parallel to the direction of motion, the x axis. The subtlety comes in the nature of the observer of the rod in frame S. The rod is observed by a Lorentz observer, which is a whole set of observers spread throughout the frame—all with synchronized clocks. At a set time, the observers all simultaneously observe the point in S' right at their point in S. Two of the observers see ends of the rod. They subsequently measure a length L by finding the distance between them.

Note that no single observer (at the origin, say) sees the same length (except in very special symmetric circumstances). If a single observer looks out at the rod, he sees the end further away from him at an earlier time than he sees the end closer to him, because it takes light longer to reach the observer from the further end.

The Lorentz observer (the whole set) sees a length L. How can we find that length from the Lorentz transformation? The trick is to keep straight exactly which positions and exactly which times are known before calculation. In the present problem, the ends of the rod, (x'_1, y'_1, z'_1) and (x'_2, y'_2, z'_2), are known in the frame S', whereas the only time known is the time t for the Lorentz observer in frame S. The relationship easiest to use is the one which relates primed spatial positions and unprimed times.

The directions perpendicular to the motion are unchanged. The Lorentz observed length L is

$$L = (x_2 - x_1) = \frac{x'_2 - x'_1}{\gamma} = L_0 \sqrt{1 - \frac{v^2}{c^2}}$$

This is the Lorentz contraction. It is the length seen by a Lorentz observer in one frame observing a length fixed in the second frame. The observers in S', the rest frame of the rod, will say that the ends were measured at *different* times t'. The times (to a Lorentz observer in the rest frame of the rod) are

$$t'_2 = \frac{t}{\gamma} - \frac{v x'_2}{c^2}$$

$$t'_1 = \frac{t}{\gamma} - \frac{v x'_1}{c^2}$$

For practice, write a short program [like LOROBS, Programs 7.1(a), (b)] which calculates the Lorentz-observed positions of moving points. Print out the times of measurements in the rest frame, too.

Program 7.1(a)

```
LOROBSBA    Lorentz relativistic observer

10 LET C=1    Speed of light
15 LET V=.9*C    Relative speed of systems
20 LET G=1/SQR(1-V*V/(C*C))    Gamma
25 INPUT T    Observation time
30 FOR I=1 TO 6
35 READ X1,Y1,Z1                       } Coordinates
40 DATA -1,-1,0,1,-1,0,1,0,0           } of object
45 DATA 1,1,0,-1,1,0,-1,0,0            } points
50 LET T1=T/G-V*X1/(C*C)    Time in object frame
55 LET X=G*(X1+V*T1)    Lorentz-transformed x
60 PRINT X;Y1;Z1
65 NEXT I
70 RESET
75 GOTO 25    Return for new observation time
80 END
```

Program 7.1(b)

```
LOROBSFO    Lorentz relativistic observer

      DIMENSION X(6,3)
      DATA C,V/1.,.9/   Speed of light, relative speed of systems
      GAM=1./SQRT(1.-V**2/C**2)    Gamma
      DATA X(1,1),X(1,2),X(1,3)/-1.,-1.,0./   )
      DATA X(2,1),X(2,2),X(2,3)/1.,-1.,0./    |
      DATA X(3,1),X(3,2),X(3,3)/1.,0.,0./     | Coordinates
      DATA X(4,1),X(4,2),X(4,3)/1.,1.,0./     > of object
      DATA X(5,1),X(5,2),X(5,3)/-1.,1.,0./    | points
      DATA X(6,1),X(6,2),X(6,3)/-1.,0.,0./    )
    5 READ(5,100)T    Observation time
      IF (T-999.) 7,20,20
    7 DO 10 I=1,6
      TPRIM=T/GAM-V*X(I,1)/C**2    Time in object frame
      XOBS=GAM*(X(I,1)+V*TPRIM)    Lorentz-transformed x
   10 PRINT 101,XOBS,X(I,2),X(I,3)
      GOTO 5    Return for new time
  100 FORMAT(F10.4)
  101 FORMAT(3E12.4)
   20 STOP
      END
```

7.4 The Speed-of-Light Distortion

The Lorentz-contracted rod was viewed by a Lorentz observer. What would a single observer placed at, say, the origin of a frame see as time progressed? If the object moves from far out $-x$ to far out $+x$, the rod will appear very much longer than its rest length at the beginning and will appear very much shorter than its Lorentz-contracted length at the finish.

The light reaching the eye comes from different points on the rod at different times for two reasons: First, because of the Lorentz transformation itself (the time effect discussed above); second, because it takes different amounts of time for light from various parts of an object to reach the eye. If the object were not moving, the second distortion would not matter; neither would the first distortion. Since the object is moving, the eye sees different parts of the object where they were at different times in the eye's own frame. This speed-of-light distortion lies entirely in the observer's own frame. The phenomenon would be observed, even without a relativistic effect. The effect is related to retarded-field effects and is only important at high velocities.

If you call t the time at which the light reaches the eye, then to include the speed-of-light distortion as well as the Lorentz transformation you must solve (for each point on the object) the equations

$$t-\sqrt{\frac{x^2+y^2+z^2}{c}}=\gamma\left(t'+\frac{vx'}{c^2}\right)$$

$$x=\gamma(x'+vt')$$

$$y=y'$$

$$z=z'$$

In these equations, the object is at rest in S' and the eye is at the origin of S. You know the time t for the eye and the coordinates (x', y', z') of a point on the object in its frame. You want to find the apparent coordinates (x, y, z) and the time t' in the object's frame. The solution reduces to a quadratic equation for t'.

$$c^2(t')^2 - 2t\gamma c^2(t') + [(c^2t^2 - 2t\gamma vx' - (x')^2 - (y')^2 - (z')^2] = 0$$

Program 7.2(a)

```
LOCOBSBA     Local relativistic observer

100 LET C=1      Speed of light
110 LET V=.9*C     Relative speed of systems
120 LET G=1/SQR(1-V*V/(C*C))     Gamma
130 PRINT "TIME THE EYE LOOKS?"
140 INPUT T
150 FOR I=1 TO 6     Step through points on object
160 READ X1,Y1,Z1                        } Coordinates
170 DATA -1,-1,0,1,-1,0,1,0,0            } of each
180 DATA 1,1,0,-1,1,0,-1,0,0             } object point
190 LET A0=1/(G*G)                                  } Coefficients
200 LET B0=-2*(X1/G+V*T)                            } for quadratic
210 LET C0=V↑2*T↑2+X1↑2/G↑2+2*X1*V*T/G              } formula
220 LET C0=C0-V↑2*(Y1↑2+Z1↑2)/C↑2                   }
230 LET X=(-B0-SQR(B0*B0-4*A0*C0))/(2*A0)     Observed
240 PRINT X;Y1;Z1                             x coordinate
250 NEXT I
260 RESET
270 GOTO 130     Return for new time
280 END
```

Program 7.2(b)

```
LOCOBSFO     Local relativistic observer

          DIMENSION X(6,3)
          DATA C,V/1.,.9/     Speed of light, relative speed of systems
          GAM=1./SQRT(1.-V**2/C**2)     Gamma
          ACOE=1./(GAM*GAM)     Quadratic coefficient
          DATA X(1,1),X(1,2),X(1,3)/-1.,-1.,0./    }
          DATA X(2,1),X(2,2),X(2,3)/1.,-1.,0./     } Coordinates
          DATA X(3,1),X(3,2),X(3,3)/1.,0.,0./      } of object
          DATA X(4,1),X(4,2),X(4,3)/1.,1.,0./      } points
          DATA X(5,1),X(5,2),X(5,3)/-1.,1.,0./     }
          DATA X(6,1),X(6,2),X(6,3)/-1.,0.,0./     }
 5        READ(5,100)T     Observation time
          IF (T-999.) 7,20,20
 7        DO 10 I=1,6     Step through object points
          BCOE=-2.*(X(I,1)/GAM+V*T)                                    } Coefficients
          CCOE=V*V*T*T+(X(I,1)*X(I,1))/(GAM*GAM)+2.*X(I,1)*V*T/GAM     } for quadratic
          CCOE=CCOE-(V*V)*((X(I,2)*X(I,2))+(X(I,3)*X(I,3)))/(C*C)      } formula
          XOBS=(-BCOE-SQRT((BCOE*BCOE)-4.*ACOE*CCOE))/(2.*ACOE)     Observed x
 10       PRINT 101,XOBS,X(I,2),X(I,3)                              coordinate
          GOTO 5     Return for new time
 20       STOP
 100      FORMAT(F10.4)
 101      FORMAT(1X,3F10.4)
          END
```

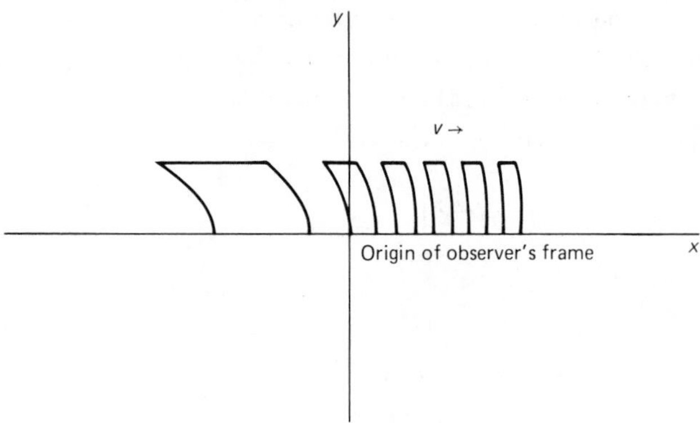

Figure 7.1 The appearance of a square moving by the observer at $0.9c$ toward $+x$. Both relativistic effects (the Lorentz transformation) and the speed-of-light distortions in the observer's own frame are taken into account. In its rest frame, the square has corners (-1, 0, 1), (0, 0, 1), (0, 1, 1), and (-1, 1, 1). The observer's times are 0.5, 1.5, . . . , 5.5 at the origin of the frame. Time is zero when the origins coincide. The figure is related to the boxcar effect.

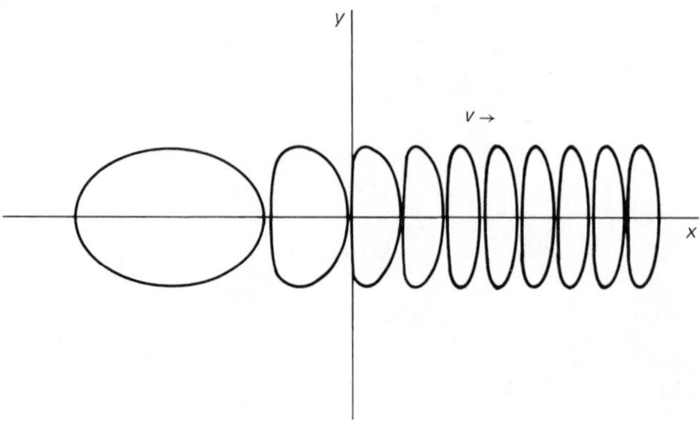

Figure 7.2 The appearance of a circle moving at $0.9c$ toward $+x$. In its rest frame the circle has radius 1 and center at (0, 0, 2). The observer is at the origin, and the times are 0.5, 1.5, . . . , 9.5.

This is solved by the quadratic formula and then substituted back in to get the coordinate x. The procedure appears to be more complicated than it is, especially when using the computer. Programs like LOCOBS [Programs 7.2(a), (b)] carry out the calculations.

Figure 7.1 shows a square as it whips by the eye. Figure 7.2 shows a circle. Using this short program you can illustrate the so-called boxcar effect yourself; choose points on the corners of a cube. You can also discover what happens to longitude and latitude lines on a moving sphere. These problems have been discussed in G. D. Scott and H. J. van Driel, *Am. J. Physics,* **38**,971 (1970), as well as in the other articles referenced there.

7.5 Electric Fields to a Moving Observer

Magnetism can be viewed as an electric field seen by a moving observer (a test charge). The completely correct way of discussing such a fact uses the **E** and **B** transformations (which are complicated and beyond the scope of this book). You can gain an understanding of what happens in the following way. Consider the field-line pattern of a single point charge as viewed by a Lorentz observer moving relative to the charge. Draw the field-line pattern in the rest frame of the charge just as you may have already done with the program *EV* from Chapter 4. At each point on a field line (or equipotential), transform the coordinates to those observed by a Lorentz observer moving parallel to x at velocity v. In this situation, you know the position in one frame and the (Lorentz observer's) time in the other, so the problem reduces to a Lorentz contraction of the x coordinates. Figure 7.3 shows the field lines and one equipotential as seen by an observer moving at $0.5c$. This is also the way a charge looks moving by the observer at $0.5c$. In the rest frame of the charge the field lines are equally spaced in angle, the equipotential is circular, and the **E** and V contours meet at right angles. When moving by at $0.5c$, the lines are packed more closely around the directions perpendicular to the motion (although they are still radial). The equipotential is elliptical, and the equipotential no longer meets the field lines at right angles.

These features are clear in Figure 7.4, wherein the patterns $v = 0.9c$ and $v = 0.99c$ are shown. In all the figures, the field lines are traced out to the same distance from the charge in its rest frame.

The elliptical contour is the path along which a charge should be moved in order to do no work on it. This contour is circular in the rest frame of the charge but elliptical in a moving frame. Even in the moving frame, the contour still represents the zero work path. Thus, the fact that this path is not perpendicular to the observed electric field lines (and hence to the electric field at each point) implies that another field exists which somehow cancels the tangential part of the **E** field. This field is wholly due to the fact that the Lorentz observer is moving relative to the field pattern. This additional field is the magnetic field **B**.

Similar field pattern changes occur for other charge distributions. Figure 7.5 shows the field patterns for a dipole moving perpendicular to its length at $v = 0.2c$ and $v = 0.9c$. Figure 7.6 shows the field pattern for a dipole moving parallel to its length at $v = 0.2c$ and $v = 0.9c$. All these patterns are those discussed in Chapter 4 using the

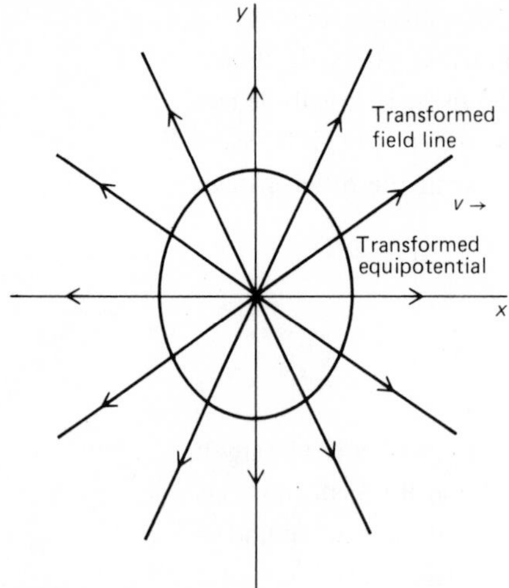

Figure 7.3 Electric field lines (and an equipotential contour) for a single point charge as seen by a Lorentz observer moving by in the x direction at $0.5c$.

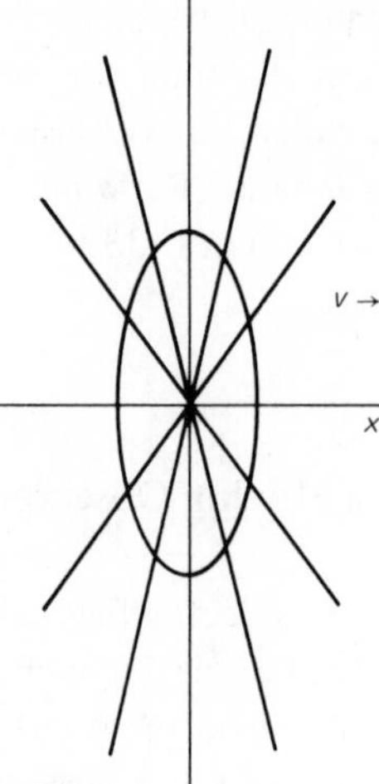

Figure 7.4 As seen by a Lorentz observer moving by in the $+x$ direction at $0.9c$. The fact that the equipotential contour (a zero work path) is not perpendicular to the field lines shows the apparent existence of another force (the magnetic force).

program EV but with each pair of coordinates Lorentz-transformed (Lorentz-contracted along x) into a moving frame of reference.

7.6 Situations with v Not Parallel to x

Often situations which occur in nature do not have the velocity of the observer parallel to the x axis. Sometimes the coordinate system can simply be rotated. At other times, however, there are reasons for setting the coordinate system in a particular way, and then the Lorentz transformation is more difficult. It is quite complicated to write down the Lorentz transformation for a velocity in an arbitrary direction. When using the computer, it is easier to perform two rotations with the usual Lorentz transformation in between.

The two-dimensional situation will serve to show the method. Suppose that the

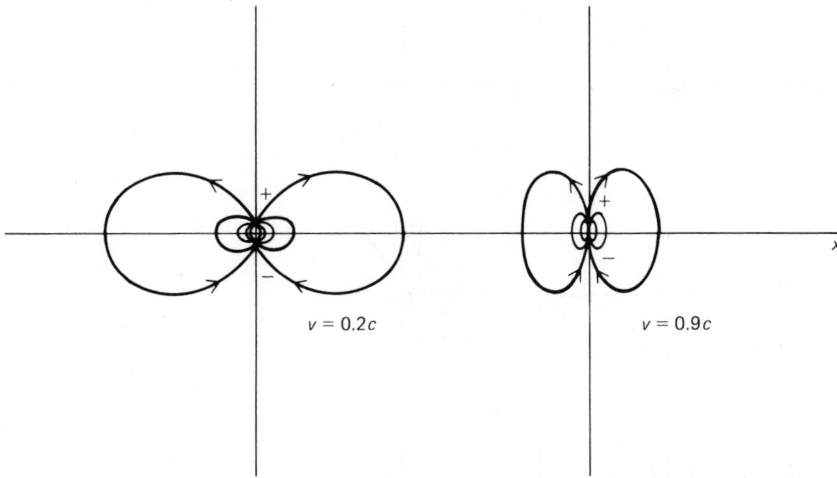

Figure 7.5 Electric-field-line patterns for a dipole as seen by a Lorentz observer moving perpendicular to the dipole with speeds of $0.2c$ and $0.9c$. The pattern is contracted parallel to the motion of the observer (or the apparent motion of the dipole). The pattern for $0.2c$ is very nearly that for a dipole at rest.

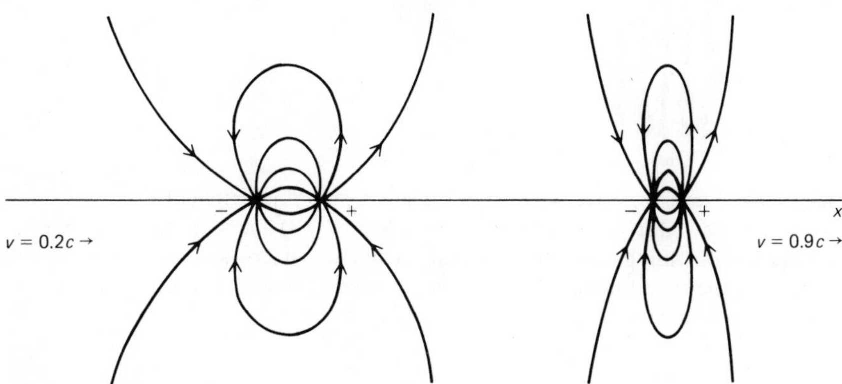

Figure 7.6 Electric-field-line patterns for a dipole moving parallel to its length at $0.2c$. Again the pattern is contracted parallel to the motion.

velocity has components v_x, v_y. First rotate the coordinates to put **v** along x. This is done by the equations

$$x_{\text{new}} = L_1 x_{\text{old}} + L_2 y_{\text{old}} \qquad y_{\text{new}} = L_3 x_{\text{old}} + L_4 y_{\text{old}}$$

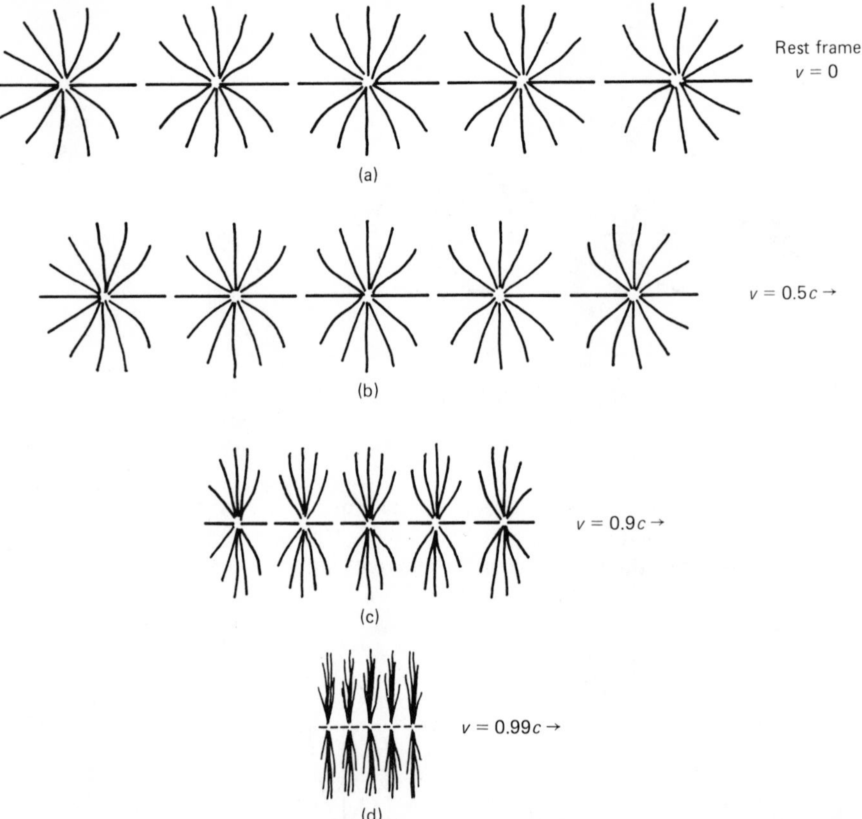

Figure 7.7 Electric-field-line patterns for five equal point charges on a line parallel to the x axis. (a) The pattern as viewed in the rest frame of the charges. (b) The pattern as seen by a Lorentz observer moving parallel to x at $0.5c$. (c) The pattern at $0.9c$. (d) The pattern at $0.99c$. All the field lines are drawn at equal angles for an equal distance out from the charges in the rest frame.

where $L_1 = L_4 = \cos Q$, $L_2 = -L_3 = \sin Q$ and Q is the angle that v makes with the old x axis. Now that the velocity is parallel to x, you can perform the usual Lorentz transformation. Then rotate the coordinate system in the new frame back through the angle Q so that the system is arranged as at the start. All these transformations can be represented by matrices, so that the series of transformations can be performed as products of matrices.

This strategy can be performed for coordinate points in S. (A three-dimensional form is not any harder but merely more tedious because the rotation making **v** parallel to x is more complicated.) Such a program calculates the position (x', y') as seen by a

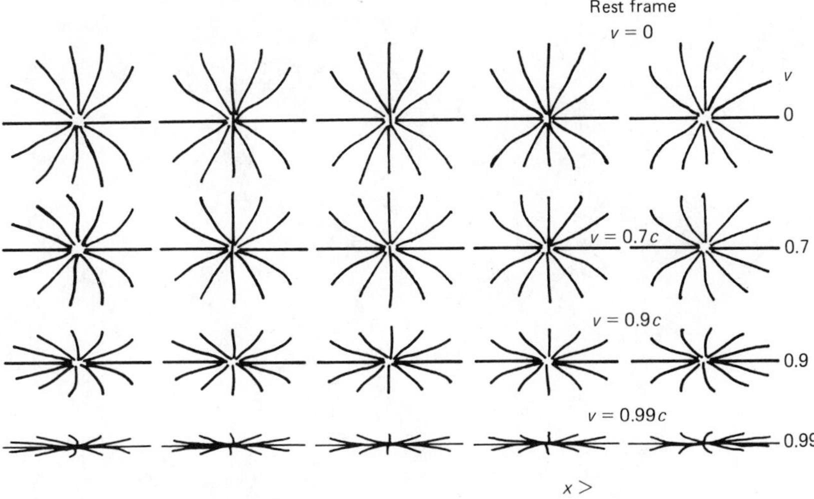

Figure 7.8 Electric field lines for five equal point charges on a line moving toward the Lorentz observer. Patterns for y speeds of 0, 0.7c, 0.9c, and 0.99c are shown. The patterns are contracted parallel to the motion of the charges (or the Lorentz observer).

Lorentz observer in the inertial frame S' given the time t' in that S' frame and the position (x, y) of the point in its rest frame S. You would also supply the components (v_x, v_y) of the velocity of S' with respect to S.

How does a strategy such as this apply to the moving charged-particle problem? Suppose you have stationary positive charges on a wire as viewed from the lab frame. A positive test charge moves near the wire with velocity (v_x, v_y). A Lorentz observer in the rest frame of the test charge sees a very different situation. He observes contracted field patterns and different charge positions for the charges. Again all you need to do is use the old field-line program but perform the rotation—Lorentz transformation—sequence on each field point calculated.

Figure 7.7(a) through (d) shows the apparent patterns for $\mathbf{v} = (0, 0)$, $\mathbf{v} = (0, 0.5c)$, $\mathbf{v} = (0, 0.9c)$ and $\mathbf{v} = (0, 0.99c)$ with $\mathbf{v}$ parallel to a line of charges. Figure 7.8 shows the patterns when v is perpendicular to the line of charges. Figure 7.9 shows the patterns when $|\mathbf{v}| = 0, 0.7c, 0.9c, 0.99c$, but $\mathbf{v}$ points at 45° to the x axis. In all the cases, the field pattern is contracted along the direction of motion of the charges. Interesting rotational effects (hooked field lines) also occur. One of the most straightforward ways to consider magnetic fields is to compare the field patterns to the ones in the rest frame of the charges (the lab frame). The difference is the part of the force due solely to the fact that the test charge is moving.

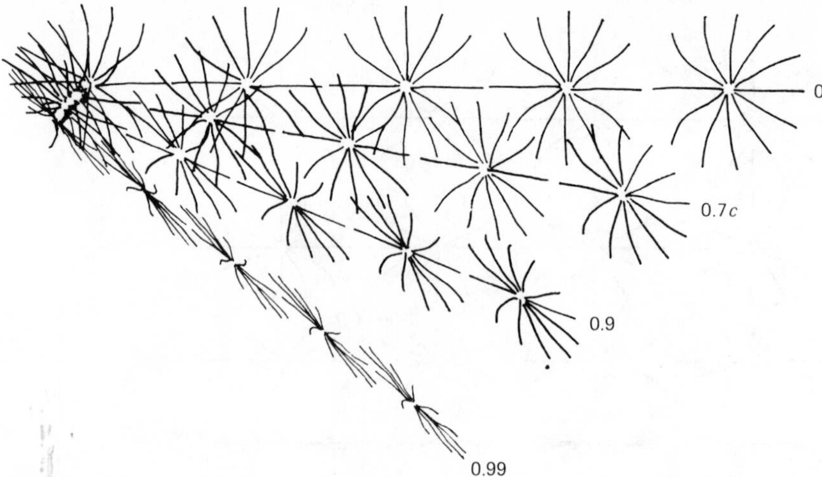

Figure 7.9 Electric-field-line patterns for the five equal point charges on a line as seen by a Lorentz observer moving at 45° to the line. Speeds of 0, 0.5c, 0.9c, and 0.99c are shown; the direction of v is always at 45° to the x axis. The field patterns are contracted in the directions perpendicular to the velocity, but the Lorentz observer sees different points at different times (according to the charge). The apparent line of charges lies at an angle to the axes.

Now consider a wire carrying a current and a moving test charge nearby. The positive charges are stationary in the lab frame. The field patterns of the moving test charge are as discussed above. The negative charges are moving in the lab frame so that the distortions of their field pattern as viewed by the test charge in its rest frame are slightly different. For the negative charges you must transform from the lab frame to their rest frame and then again from their rest frame to the rest frame of the test charge. This difference between the + and the - electric fields as viewed by a Lorentz observer in the rest frame of the test charge is considered in the lab frame as the magnetic field. Using only the simple programming you have now seen, you can calculate the electric field and magnetic field correspondences.

Problems

Problems in relativity are often not as useful as the pictures which can be produced from results of programs. The following are exercises which might prove useful.

1. a. Write a program which performs a Lorentz contraction for lengths parallel to x but leaves the y and z directions alone. To what physical situation does this correspond?

b. Consider a sphere moving by at 99% the speed of light. By finding points on lines of longitude and latitude in the rest frame of the sphere and by using your program from part a, plot the appearance of the sphere to a Lorentz observer in your frame.

c. What other effects would you need to consider in order to plot the appearance of the sphere to a local observer in your frame?

2. a. Write a program which performs a Lorentz contraction for lengths parallel to x but leaves the y and z directions alone. To what physical situation does this correspond?

b. Consider a point charge moving by you at 99% the speed of light. By calculating points on field lines and equipotentials in the rest frame of the charge and by using your program to transform those points to your frame of reference, plot the apparent field line and equipotential pattern as seen by a Lorentz observer in your frame.

c. Explain the relationship between your plot and the concept of magnetic fields.

8 circuit theory

8.1 Introduction

In much of your everyday experience you must deal with electricity, electrical machinery, and electronic instrumentation, so the theory of circuits is a useful subject. Even though there is not time to dig deeply into the circuit theory for active elements, such as transistors or vacuum tubes, it is possible to get a feeling for passive element circuit theory and to start to understand phase shifts, resonance, and frequency dependence.

This chapter deals first with *RC* circuits, then briefly with *LR* circuits, and finally with *LRC* combinations. The initial approach taken here is the integration of differential equations (not the use of impedance), so that nonlinear circuit elements can be used. The differential equations describing voltages and currents flowing through resistors, capacitors, and inductors are reasonably simple to handle. The equations are in many ways similar to those of kinematics and dynamics. For example, the circuit for *LRC* resonances has exactly the same differential equation as the force equation for damped harmonic motion.

The chapter also deals with the concept of impedance, Kirchhoff's laws for dc circuits, bridge circuits, potentiometer circuits, and simple circuits such as the *RC* integrator and differentiator circuits.

8.2 Basic Equations

The basic equations governing the relationship between currents and voltages in resistors, capacitors, and inductors are derived in your text. If something here looks new, review your text. Most of the equations amount to a combination of a physical fact and a definition. Ohm's law, for instance, states the physical fact that for many materials (such as pieces of wire), the potential difference (voltage) across the element is directly proportional to the current going through the element. Ohm's law also defines a quantity called the resistance R as the coefficient of proportionality. R is still a usable quantity (as $R = V/I$) even if V is not proportional to I. A brief resume of the relations you will use in this chapter follows.

1. $C = \dfrac{q}{V}$ *(Definition of capacitance)*

   where q = charge on each plate of the capacitor
   V = voltage across the capacitor

2. $I = -\dfrac{dq}{dt}$

   Current is the time rate of change of the charge on the capacitor.

3. $V = RI$ *(Ohm's law)*

4. $P = I^2 R$

   $P = V^2/R$ *(Power dissipation in a resistor)*

5. $\mathcal{E} = -L\dfrac{dI}{dt}$ *(Definition of self-inductance)*

   where $\mathcal{E}$ = emf developed across a coil having self-inductance L
   I = current through the coil

6. $V = L\dfrac{dI}{dt}$

   where V = voltage drop *across* an inductor
   I = current through the inductor

7. $\mathcal{E} = -M\dfrac{dI}{dt}$ *(Definition of mutual inductance)*

   where I = current in one coil
   $\mathcal{E}$ = emf produced in the other coil
   M = mutual inductance between the two coils

8. Energy stored = $\begin{cases} \frac{1}{2} CV^2 & \text{in a capacitor} \\ \frac{1}{2} LI^2 & \text{in an inductor} \end{cases}$

9. $R_{\text{series}} = R_1 + R_2 \qquad R_{\text{parallel}} = \dfrac{1}{1/R_1 + 1/R_2}$

$L_{\text{series}} = L_1 + L_2 \qquad L_{\text{parallel}} = \dfrac{1}{1/L_1 + 1/L_2}$

$C_{\text{series}} = \dfrac{1}{1/C_1 + 1/C_2} \qquad C_{\text{parallel}} = C_1 + C_2$

10. $C = \mathcal{E}_0 \dfrac{A}{d}$ *(Parallel plates)*

where C = capacitance of capacitors
A = area of each plate
d = separation

11. Kirchhoff's laws: (1) the (algebraic) sum of all currents entering a junction (node) is zero. (2) The (algebraic) sum of all voltage drops around a loop (circuit) is zero.

8.3 Circuits with Resistors and Capacitors

Consider a simple physical situation in which an initially charged capacitor C is connected across a resistor R at time $t = 0$ (Figure 8.1). The voltage elements are equal, since carrying a charge from A to B by either arm changes the electric potential of the charge by the same amount. V across $C = V$ across R. Using basic equations 1 and 3 above,

$$\frac{q}{C} = RI \quad \text{or} \quad \frac{dq}{dt} = -\frac{q}{RC}$$

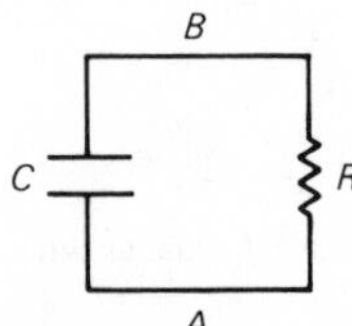

Figure 8.1 An initially charged capacitor C connected across a resistor R.

The differential equation has a simple closed-form solution, namely, $q = q_{\text{initial}} \times \exp(-t/RC)$. However, for the moment, let us integrate the equation numerically. Given the present value of the charge q_0 on the capacitor, calculate the current $I(= -dq/dt)$. Now find the new charge q a moment Δt later as $q = q_0 - I\Delta t$. Repeat.

By writing and executing your own such program (with $R = 10^6\ \Omega$, $C = 10^{-6}$ F, and $q_{\text{initial}} = 10^{-6}$ C), show that an exponential decay of the charge to zero occurs. What is the time dependence of the current I? What was the current before $t = 0$? (Assume the capacitor has always been charged for $t < 0$.)

The calculation just discussed was almost trivial. As an exercise, let us now try something more interesting. Resistors are practically always temperature-dependent. You can easily arrange for the temperature to rise in a resistor because there is I^2R heating due to any current flowing. Incandescent light bulbs shine because the filament is heated to white hot temperatures by I^2R heating. The resistance of light bulbs often changes substantially when the current is turned on; a 75-W bulb goes from $R = 20\ \Omega$ at no current to $R = 200\ \Omega$ at full current. The resistance of a light bulb depends on I^2 for small I.

Suppose that the resistance is given by

$$R = (10^6)(1 + 0.1)(I^2)$$

Now find the charge and current as a function of time. This problem is still simple on the computer, but it can not be solved in closed form. While you are about it, calculate the total energy dissipated (up to time t) as I^2R power:

$$\text{energy} = \frac{1}{t}\int_0^t I^2 R\, dt$$

Now let us add a battery. The circuit diagram is shown in Figure 8.2; the differential equation is

$$\frac{q}{C} = RI + E_b$$

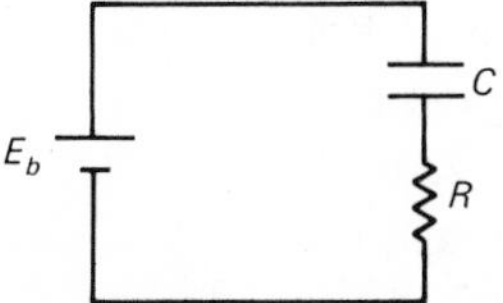

Figure 8.2 A battery E_b connected across a series resistor R and capacitor C.

Program 8.1(a)

```
RCBAS    RC decay (nonlinear resistor)

100 LET Q=0          )
110 LET T=0          |
120 LET R0=1E4       |
130 LET R=R0         } Initialization
140 LET V=10         |
150 LET C=1E-6       )
160 LET D=R*C/10      Time step, Δt
170 LET I=Q/(R*C)-V/R      Initial current
180 PRINT "RC=";R*C;" SEC.  Q(0)=";Q;" COUL."
190 PRINT "V(0)=";V;" V FINAL CHARGE=";C*V;" COUL."
200 PRINT "  TIME","  CHARGE","  CURRENT","  RESISTANCE"
210 PRINT T,Q,I,R
220 LET R=R0*(1+1E4*I*I)     New resistance
230 LET I=Q/(R*C)-V/R     New current
240 LET Q=Q-I*D     New charge
250 LET T=T+D     New time
260 PRINT T,Q,I,R
270 IF T<2*R*C THEN 220     Go back for next Δt step
280 END
```

Program 8.1(b)

```
RCFOR    RC decay (nonlinear resistor)

         Q=0.           )
         T=0.           |
         R0=1.E4        |
         R=1.E4         } Initialization
         V=10.          |
         C=1.E-6        )
         DELT=R*C/10.     Time step, Δt
         AI=Q/(R*C)-V/R     Initial charge
         PRINT 100,T,Q,AI,R
10       R=R0*(1.+1.E4*AI*AI)     New resistance
         AI=Q/(R*C)-V/R     New current
         Q=Q-AI*DELT     New charge
         T=T+DELT     New time
         PRINT 100,T,Q,AI,R
         IF (T-2.*R*C) 10,20,20     Return for next Δt step
20       STOP
100      FORMAT(1X,4(2X,E12.6))
         END
```

Let $q_{initial} = 0$ and calculate the time dependence of the charge on the capacitor. You will still find an exponential time dependence. Again the current will jump at $t = 0$ (as the switch is closed) and then die to zero. Now the charge q will start at zero and grow asymptotically to a maximum value. At q_{max}, the voltage across the capacitor matches the voltage across the battery and no current results. This is how you charge a capacitor. Again you can calculate power dissipation and find out what happens with variable resistances. Program RC [Programs 8.1(a), (b)] carries out these calculations.

Another interesting problem occurs when the battery is replaced by a source of alternating voltage. This source might be a wall outlet. (House current is a sine wave varying at 60 cycles per second, or 60 Hz.) The source might also be an audio oscillator in a physics lab, in which case you can change the frequency of the sine wave. Draw the circuit diagram (the symbol for an alternating voltage is ⊙) and write down the differential equation. The differential equation looks very much like the one given above except that the battery emf E_b is now E_0 sin (ωt).

Rewrite your program for an ac voltage of amplitude $E_0 = 1$ V and $\omega = 2\pi$ (1 Hz). Your algorithm to solve the equation is still okay. Execute the program and find the current (I), the voltage across the capacitor ($V = q/C$), and the voltage across the generator ($E_0 \sin \omega t$) as functions of time. Start the system with $q = I = t = 0$. Plot the results as a function of time. Let several cycles pass so that the circuit settles down. The current through the circuit does not pass through zero when the driving voltage ($E_0 \sin \omega t$) does. This effect is called a "phase shift" between I and V.

The current could be written as $I = I_0 \sin (\omega t + \theta)$, where θ represents the time difference between zero crossing (Δt) written as an angle ($\theta = \omega \Delta t$). This phase shift is a function of frequency, as is the maximum current which flows. Run the program for $\omega = 2\pi/100$ and $\omega = 100\ (2\pi)$. See if the phase shifts are different. At the lowest frequency, the capacitor dominates the circuit; at the highest, the resistor dominates. See if you can figure out what phases ought to occur at those extreme frequencies. The total instantaneous power dissipated is VI, where V is the total voltage across the circuit. Modify your program to calculate the power dissipated at each time step and the average power dissipated up to that time. Do these numbers make any physical sense to you?

8.4 Circuits with Inductors and Resistors

Circuits with self-inductors L and resistors R are in many ways similar to circuits with capacitors C and resistors R. The phase shifts generated by ac voltages are usually of the opposite sign, that is, currents in capacitive circuits lead the voltage, whereas currents in inductive circuits lag. The differential equation for the series circuit of an ac voltage generator, an inductor L, and a resistor R is

$$E_0 \sin (\omega t) = L \frac{dI}{dt} + RI$$

Write a program to solve this equation algorithmically for the current in the circuit and the various voltages. Again look for phase shifts as a function of frequency. Typical values might be $L = 10^{-3}$ H, $R = 10^3\ \Omega$, $E_0 = 1$ V, and $\omega = 2\pi(10^6)$ rad/s (an AM radio frequency). (You might find it easier to consider the dc battery case first.)

8.5 Circuits with Inductors, Capacitors, and Resistors

Capacitors and inductors both store energy. Capacitors store energy in electric fields; inductors store energy in magnetic fields. When both capacitors and inductors are combined in a single circuit, it is possible to trade energy back and forth. At one point in time, the energy might be mostly stored in the electric field of the capacitor; at some later moment, the energy might be mostly stored in the magnetic field of the inductor. This interchange of energy leads to oscillations, and these oscillations have important applications in electronics. In radio and television systems, oscillations in *LC* circuits (tuned circuits or tank circuits) are used to select one station out of all the signals in the air.

The frequency with which the energy is transferred back and forth between the capacitor and the inductor depends on the values of L and C. This natural resonance frequency, $\omega_r = 1/\sqrt{LC}$, is a property of the circuit. If you put a signal with a different frequency (from an oscillator, for example) into the circuit, you might expect interesting effects. When the two frequencies are equal, you are trying to make the system respond at its natural frequency. Usually the system likes to respond at its natural frequency, and spectacular results occur. This phenomenon of "resonance" is found in many fields of physics and engineering. It is an important general property to understand.

Let us consider an algorithmic approach to the currents and voltages in an *LCR* circuit. We might just as well include the resistance, because real inductors are coils of wire and so always have a resistance R as well as an inductance L.

The differential equation for a circuit including an initially charged capacitor C across a series combination of an inductor L and a resistor R is

$$\frac{q}{C} = L\frac{dI}{dt} + RI$$

This equation involves the three related quantities $I, dI/dt$, and $q = \int I\,dt$. The calculation is straightforward. You initialize the variables and then calculate

$$dI = (q/C - RI)\,dt/L.$$

This dI gives the new current $I = I + dI$ and the new charge $q = q + I\,dt$. (For more accuracy, you can calculate the current a half-step ahead of the present charge.) Write and run your own program to calculate the current and voltages in this series *LRC* circuit. Use $L = 10^{-3}$ H, $R = 1\ \Omega$, and $C = 0.1\ \mu$F. From executions of your program, find the frequency of natural resonance for the system and compare it with $\omega = 1/\sqrt{LC}$.

The really interesting case of *LRC* circuit resonance includes a sinusoidal signal generator. Draw the circuit diagram for the driven series *LRC* circuit and write down

Program 8.2(a)

```
LRCBAS    Driven LRC circuit

100 LET V0=1     Drive amplitude                        ┐
110 LET L=10E-3     Inductance                          │
120 LET R=10     Resistance                             │
130 LET C=1E-8     Capacitance                          │
140 LET W0= 1E5     Angular frequency                   │
150 LET V=V0*SIN(W0*T)     Initial drive voltage        ├Initialization
160 LET Q=0     Initial charge                          │
170 LET D=1/(20*W0)     Time step, Δt                   │
180 LET D1=1/(2*W0)     Print time                      │
190 LET T0=10/W0     Final time                         ┘
200 PRINT "W0=";W0;" RAD/SEC.;    V0=";V0;" VOLTS."
210 LET W9=1/SQR(L*C)
220 PRINT "SQR(1/LC)=";W9;" RAD/SEC.;   Q=";W9*L/R
230 PRINT "TIME STEP=";D;" SEC.; W0*TIME STEP=";W0*D
240 PRINT
250 PRINT "TIME","CURRENT","DRIVE V"
260 PRINT T,I,V
270 LET V=V0*SIN(W0*T)     Drive voltage
280 LET I1=(V-Q/C-R*I)/L     dI/dt
290 IF T>0 THEN 320 ┐
300 LET I=I+I1*D/2  ├Initial half-step
310 GOTO 340        ┘
320 LET I=I+I1*D     New current, I
330 LET Q=Q+I*D     New charge
340 LET T=T+D     New time
350 IF T<D1 THEN 380                   ┐
360 LET D1=D1+1/(2*W0)                 ├Print group
370 PRINT T-D/2,I,V0*SIN(W0*(T-D/2))   ┘
380 IF T<T0 THEN 270     Return for next Δt step
390 END
```

Program 8.2(b)

```
LRCFOR    Driven LRC circuit

        VAMP=1.     Drive amplitude                 ┐
        AL=1.E-3     Inductance                     │
        R=10.     Resistance                        │
        C=1.E-8     Capacitance                     │
        W0=1.E5     Angular frequency               │
        DELT=1./(20*W0)     Time step, Δt           ├Initialization
        PRNTT=10.*DELT     Print time               │
        ENDT=10./W0     Final time                  │
        T=0.                                        │
        AI=0.                                       ┘
5       V=VAMP*SIN(W0*T)     Drive voltage
        DELI=(V-Q/C-R*I)/AL     dI/dt
        IF(T) 10,10,20             ┐
10      AI=AI+DELI*DELT/2.         ├Initial half-step
        GOTO 30                    ┘
20      AI=AI+DELI*DELT/2.     New current, I
30      Q=Q+AI*DELT     New charge
        T=T+DELT     New time
        IF(T-PRNTT) 50,40,40       ┐
40      PRNTT=PRNTT+(10*DELT)      │
        T1=T-DELT/2.               ├Print group
        V1=VAMP*SIN(W0*T1)         │
        PRINT 100,T1,AI,V1         ┘
50      IF (T-ENDT) 5,60,60     Return for next Δt step
60      STOP
100     FORMAT (1X,3E12.6)
        END
```

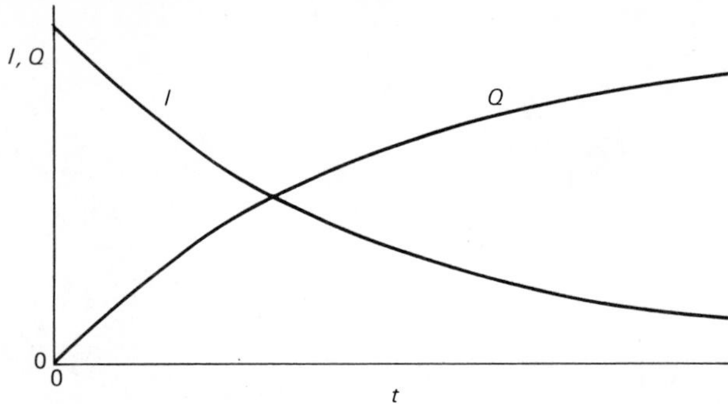

Figure 8.3 Charge and current as a function of time in an RC circuit with a battery. The charge rises exponentially to its final value.

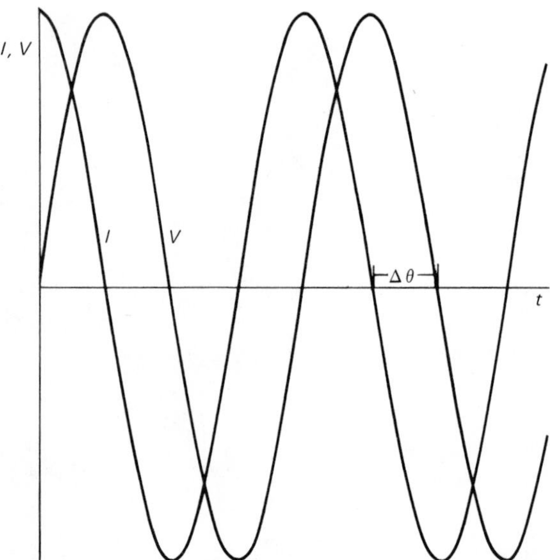

Figure 8.4 Current and drive voltage for an RC circuit driven by an audio oscillator. Notice the phase shift between the voltage and the current; the phase shift appears as a time difference between zero crossings.

the differential equation. Write a program to calculate the current and voltages in this case. Use the L, R, and C values above and make the amplitude of the drive voltage equal to 1 V with a frequency of $\omega = 10^5$ rad/s. Look for phase shifts. The program LRC [Programs 8.2(a), (b)] is a possible way to write the code. Execute the program

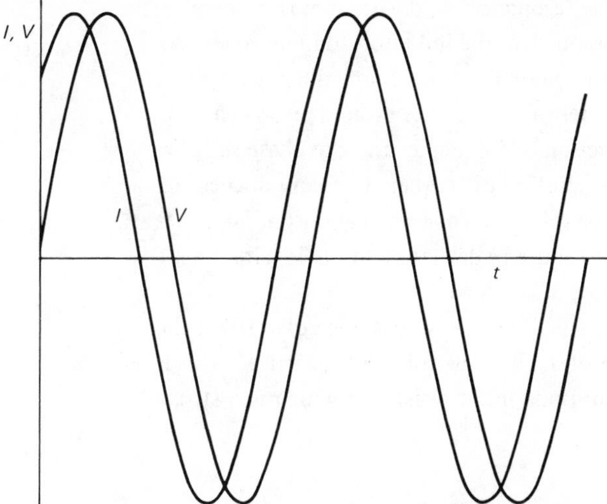

Figure 8.5 Current and voltage for a driven series *LRC* circuit. The frequency of the drive voltage is less than the resonant frequency, so the capacitor dominates the circuit. You can see the phase shift.

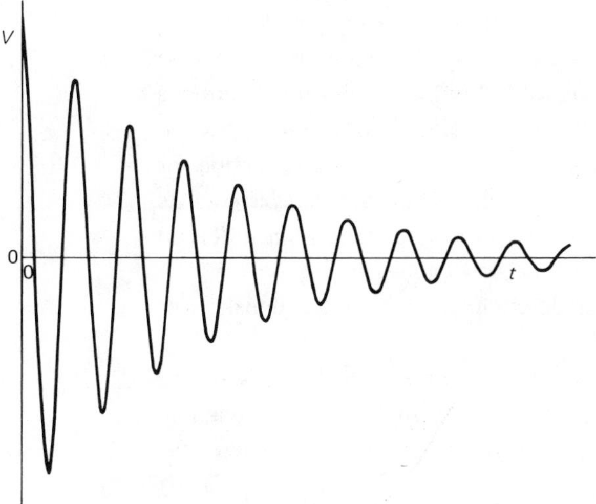

Figure 8.6 *LRC* damped oscillations (response to a step voltage over a long time period).

for frequencies well below and well above the resonance condition. Below resonance, the capacitor dominates the circuit; above resonance, the inductor dominates; at resonance, the phase effects of the capacitor and inductor cancel and only the resistor counts. By executing the program for a number of frequencies around resonance, you can calculate the maximum current as a function of frequency and the phase shift between the current and the drive voltage as a function of frequency. Many of these relationships are true for all resonance situations; resonance occurs in almost all branches of physics and engineering. Series resonant (and parallel resonant) *LRC* circuits are prototypes of resonances in general (see Figures 8.3 to 8.6).

Replace the resistor R by a diode. Assume the diode has a resistance of 1 MΩ if the voltage across it is negative, but a resistance of 1 Ω if the voltage is positive. This is how a radio detector circuit works. A real diode has a resistance which is a strong function of voltage or current.

8.6 Kirchhoff's Laws

Elementary circuit theory allows you to use some techniques of advanced BASIC programming. Kirchhoff's law calculations are a natural use of matrix manipulations. Concepts of impedance in single ac circuits illustrate complex variable calculations.

Kirchhoff's laws are statements of conservation of charge and conservation of energy. They are (I) the (algebraic) sum of the currents entering any junction is zero; and (II) the (algebraic) sum of the voltage drops around any closed loop is zero. These two statements always ensure enough equations to find unique values for all currents and voltages in any complicated circuit. For the dc case, Kirchhoff's laws are general; they are based on the linearity of circuit elements, on the continuity equation for electric charge, and on the fact that the electrostatic field is conservative. The laws are also valid for ac circuits as long as impedance (Z) is substituted for resistance (R) and as long as the changes in the currents and voltages are slow enough.

As an illustration of Kirchhoff's laws for dc circuits, consider an (unbalanced) bridge circuit, such as in Figure 8.7.

The R's are resistors, the V is a battery, the M is some meter, and the I's are the (unknown) currents. Kirchhoff's laws allow us to write down a number of relationships between the circuit elements, voltages, and currents. A number of these relationships are

Loop equations:

$$V_1 - R_1(I_1) - R_2(I_2) - R_4(I_4) = 0$$
$$V_1 - R_1(I_1) - R_3(I_3) - R_5(I_5) = 0$$
$$R_2(I_2) + R_4(I_4) - R_5(I_5) - R_3(I_3) = 0$$
$$R_2(I_2) + R_6(I_6) - R_3(I_3) = 0$$
$$R_4(I_4) - R_5(I_5) - R_6(I_6) = 0$$

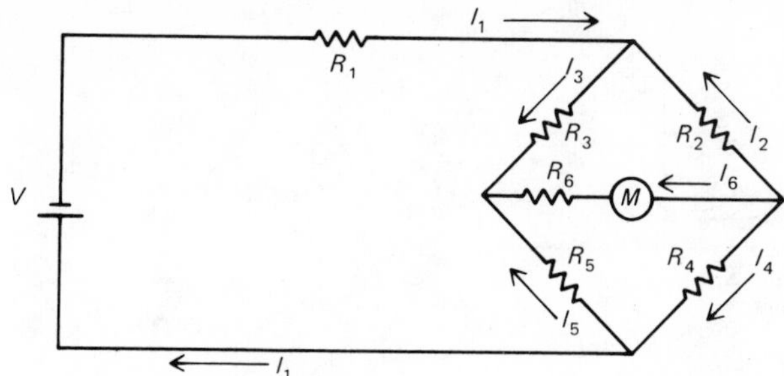

Figure 8.7 A Wheatstone bridge circuit containing the battery V, the resistors, and the currents. M is the meter which shows zero current when the bridge is balanced.

Junction equations:

$$I_1 - I_2 - I_3 = 0$$
$$I_1 - I_4 - I_5 = 0$$
$$I_2 - I_6 - I_4 = 0$$
$$I_3 + I_6 - I_5 = 0$$

Out of these (and many other similar) equations, we choose a set of six linearly independent equations for the six unknown currents. Rewriting the six chosen equations so that the unknowns line up, we have

$$\begin{array}{cccccccc}
+I_1 & -I_2 & -I_3 & & & & = & 0 \\
 & +I_2 & & -I_4 & & -I_6 & = & 0 \\
 & & +I_3 & & -I_5 & +I_6 & = & 0 \\
+R_1(I_1) & +R_2(I_2) & & +R_4(I_4) & & & = & V_1 \\
 & +R_2(I_2) & -R_3(I_3) & & & +R_6(I_6) & = & 0 \\
 & & & +R_4(I_4) & -R_5(I_5) & -R_6(I_6) & = & 0
\end{array}$$

or, using matrix notation,

$$\begin{bmatrix} 1 & -1 & -1 & 0 & 0 & 0 \\ 0 & +1 & 0 & -1 & 0 & -1 \\ 0 & 0 & +1 & 0 & -1 & +1 \\ +R_1 & +R_2 & 0 & +R_4 & 0 & 0 \\ 0 & +R_2 & -R_3 & 0 & 0 & +R_6 \\ 0 & 0 & 0 & +R_4 & -R_5 & -R_6 \end{bmatrix} \begin{bmatrix} I_1 \\ I_2 \\ I_3 \\ I_4 \\ I_5 \\ I_6 \end{bmatrix} = \begin{bmatrix} 0 \\ 0 \\ 0 \\ V_1 \\ 0 \\ 0 \end{bmatrix}$$

Program 8.3

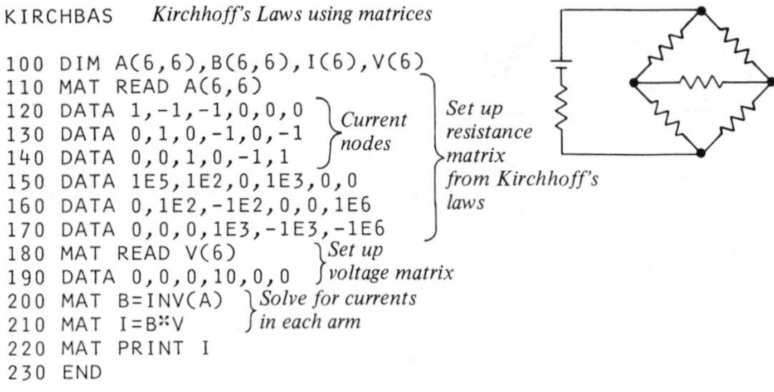

KIRCHBAS *Kirchhoff's Laws using matrices*

```
100 DIM A(6,6),B(6,6),I(6),V(6)
110 MAT READ A(6,6)
120 DATA 1,-1,-1,0,0,0
130 DATA 0,1,0,-1,0,-1
140 DATA 0,0,1,0,-1,1
150 DATA 1E5,1E2,0,1E3,0,0
160 DATA 0,1E2,-1E2,0,0,1E6
170 DATA 0,0,0,1E3,-1E3,-1E6
180 MAT READ V(6)
190 DATA 0,0,0,10,0,0
200 MAT B=INV(A)
210 MAT I=B*V
220 MAT PRINT I
230 END
```


[There is no FORTRAN program because of the differences between matrix packages on different computers.]

If you call the (known) matrix of coefficients A (a matrix with six rows and six columns) and the (known) matrix (six rows and one column) on the right side V, then the solution for the unknown matrix (six rows and one column) of currents I is

$$I = A^{-1} V$$

A^{-1}, the inverse of the matrix A, can be easily computed on many computers with BASIC compilers by the statement

```
MAT B=INV(A)
```

A complete program to solve the above bridge problem is

```
MAT READ A
MAT READ V
MAT B=INV(A)
MAT I=B*V
MAT PRINT I
```

The coefficients in the known matrices are typed in as data. (See the program KIRCH, Program 8.3.) Notice that there is no need for the bridge to be balanced. If the bridge is balanced, then $I_6 = 0$ and $I_2 = -I_4, I_3 = -I_5$. Furthermore, at balance $I_1 = [V_1/(R_1 + R_2 + R_4)]\ [(R_3 + R_5)/(R_2 + R_3 + R_4 + R_5)]$.

In general, off-balance bridges are very complicated to handle analytically. Here, however, you have an easy method. When the bridge is not balanced, current will then flow through the meter M. The current flowing from the battery will be some

Program 8.4

POTENBAS *Potentiometer*

```
100 DIM A(3,3),B(3,3),V(3),I(3)
110 LET R1=9E2
120 LET V1=20
130 LET R0=1E2
140 LET R2=1E2
150 PRINT "BATTERY TO BE MEASURED (VOLTS)"
160 INPUT V2
170 PRINT "FRACTION OF SLIDEWIRE"
180 INPUT A0
190 LET A(1,1)=R1+(1-A0)*R0
200 LET A(1,2)=0
210 LET A(2,1)=0
220 LET A(1,3)=A0*R0
230 LET A(2,3)=A0*R0
240 LET A(3,1)=-1
250 LET A(3,2)=-1
260 LET A(3,3)=+1
270 LET V(1)=V1
280 LET V(2)=V2
290 LET V(3)=0
300 MAT B=INV(A)
310 MAT I=B*V      Currents in arms
320 PRINT "CURRENTS:"
330 MAT PRINT I
340 PRINT
350 PRINT "IMPEDANCE OF POT. =";ABS((V2-R2*I(2))/I(2))
360 PRINT
370 GOTO 170      Return for new balance
380 END
```

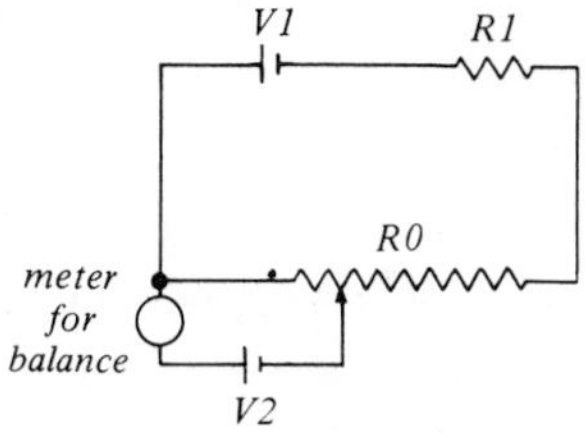

[There is no FORTRAN program because of differences in handling matrices on different computers.]

complicated function of the off-balance conditions. The current flowing through M is also a fairly complicated function.

The ratio of the voltage applied to the bridge ($V_1 - R_1 I_1$) to the current fed to the bridge (I_1) is called the input impedance. The output impedance is the effective resistance that the bridge circuit causes to the flow of current to the meter circuit. This output impedance can be found (for any given imbalance) by finding the currents and voltages for two large values of R_6. Both the input and output impedances of bridge circuits can be determined as functions of imbalance. Analytic calculations of the quantities are very complicated indeed.

Other networks can easily be studied using this Kirchhoff's laws and matrix method. Another example is the dc potentiometer. (See the program POTEN, Program 8.4.) Again the effects of imbalance are of practical importance. Even if your computer has no MAT package (as is the case with most FORTRAN languages), matrix manipulations can be performed by subroutines. The matrix inversion can be performed by a Gauss-Seidel elimination method (see B. Carnahan, H. A. Luther and J. O. Wilkes, *Applied Numerical Methods*, John Wiley & Sons, New York, 1969).

8.7 Simple ac Voltage Divider

A simple ac voltage divider shows how you can apply the computer to simple ac circuits using the concepts of impedance. A voltage divider of impedances $Z_1 = R_1 + \sqrt{-1}\, I_1$ and $Z_2 = R_2 + \sqrt{-1}\, I_2$ is drawn as shown in Figure 8.8. (Here R_1 and R_2 are the real parts of Z_1 and Z_2; I_1 and I_2 are the imaginary parts.)

$$\frac{V_{\text{out}}}{V_{\text{in}}} = \frac{Z_2}{(Z_1 + Z_2)}$$

Since Z_1 and Z_2 are complex quantities and most BASIC compilers do not include even the simplest complex number arithmetic, you will find it useful to define several functions of several variables. These functions will perform the complex arithmetic you need. If your BASIC does not include functions of more than one variable, the same operations can be performed with subroutines (GOSUB line #).

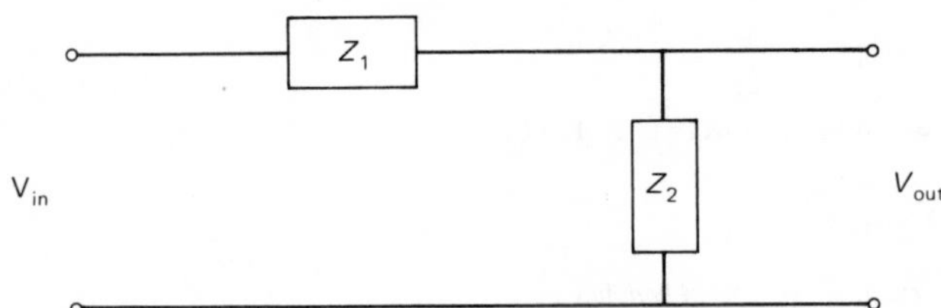

Figure 8.8 A voltage divider of two impedances Z_1 and Z_2.

For the present problem, it is convenient in BASIC to define functions for the complex operations of division (FNR and FNI), modulus or absolute value (FNM), and phase angle (FNP).

```
DEF FNR (R1,I1,R2,I2)=(R1*R2+I1*I2)/(R2*R2+I2*I2)
DEF FNI (R1,I1,R2,I2)=(R2*I1-R1*I2)/(R2*R2+I2*I2)
DEF FNM (R,I)=SQR(R*R+I*I)
DEF FNP (R,I)=180*ATN(I/R)/3.14159
```

(Most FORTRAN languages can support complex numbers directly.)

FNR and FNI are the real and imaginary parts of the quotient of two complex numbers. FNM and FNP are the modulus and the phase (in degrees) for the complex number $Z = R + \sqrt{-1}\, I$.

Using these functions you can find the amplitude (or "gain") $|V_{\text{out}}/V_{\text{in}}|$ and the relative phase of the output voltage V_{out} with respect to V_{in} for the ac voltage divider. Since the impedances Z_1, Z_2 of the circuit are practically always frequency-dependent, you might find $|V_{\text{out}}/V_{\text{in}}|$ and the phase as functions of angular frequency. The program ACDVDR [Programs 8.5(a), (b)] shows you one method for treating the simple voltage divider of Figure 8.9. The voltage divider circuit is often used in electronics and is called either an "integrator" or a "low-pass filter," depending on its use.

The complex-variable strategy is applicable to much more complicated circuits. Some BASIC compilers (and most FORTRAN compilers) can even handle complex matrix operations. On these systems, you can do Kirchhoff's law problems for ac networks.

Program 8.5(a)

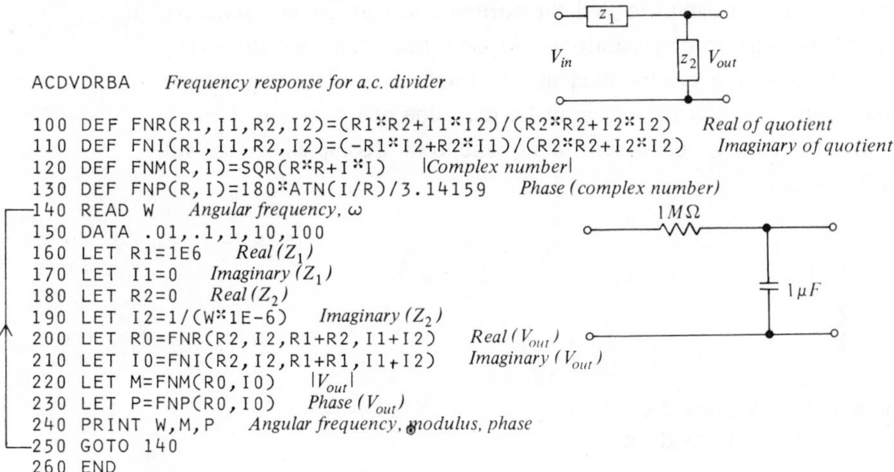

ACDVDRBA *Frequency response for a.c. divider*

```
100 DEF FNR(R1,I1,R2,I2)=(R1*R2+I1*I2)/(R2*R2+I2*I2)   Real of quotient
110 DEF FNI(R1,I1,R2,I2)=(-R1*I2+R2*I1)/(R2*R2+I2*I2)   Imaginary of quotient
120 DEF FNM(R,I)=SQR(R*R+I*I)   |Complex number|
130 DEF FNP(R,I)=180*ATN(I/R)/3.14159   Phase (complex number)
140 READ W   Angular frequency, ω
150 DATA .01,.1,1,10,100
160 LET R1=1E6   Real (Z1)
170 LET I1=0   Imaginary (Z1)
180 LET R2=0   Real (Z2)
190 LET I2=1/(W*1E-6)   Imaginary (Z2)
200 LET R0=FNR(R2,I2,R1+R2,I1+I2)   Real (Vout)
210 LET I0=FNI(R2,I2,R1+R1,I1+I2)   Imaginary (Vout)
220 LET M=FNM(R0,I0)   |Vout|
230 LET P=FNP(R0,I0)   Phase (Vout)
240 PRINT W,M,P   Angular frequency, modulus, phase
250 GOTO 140
260 END
```


Program 8.5(b)

ACDVDRFO *Frequency response of a.c. divider*

```
      COMPLEX V,Z1,Z2
10    READ(5,100)OMEGA
      IF (OMEGA-999.) 20,30,20
20    Z1=(1.E6,0.)
      Z2=CMPLX(0.,1./(OMEGA*1.E-6))
      V=Z2/(Z1+Z2)   Vout
      AM=CABS(V)   |Vout|
      PHAS=180.*ATAN2(AIMAG(V),REAL(V))/3.14159   Phase (Vout)
      PRINT 101,OMEGA,AM,PHAS
      GOTO 10
30    STOP
100   FORMAT (F10.5)
101   FORMAT (1X,3E10.4)
      END
```

Z1, Vin, Z2, Vout

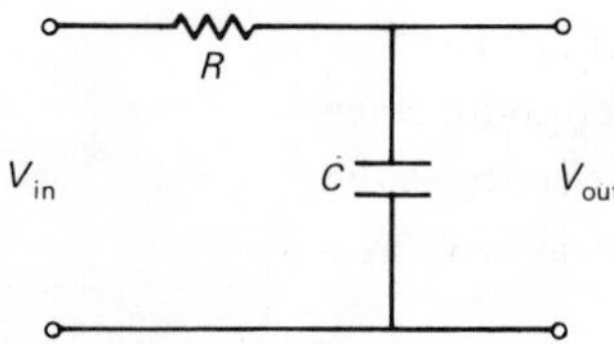

Figure 8.9 One example of a voltage divider also known as an integrator or a low-pass filter.

Problems

Many of the problems which can be done directly with the program approach have already been illustrated by questions in the text. Just a few more will be mentioned here as illustrations of what you can do.

1. Write and execute a program to find the current and voltages in a series RC circuit (with no battery or generator). At each time step, find the energy stored in the capacitor's electric field and the total energy so far expended in I^2R dissipation (Figure 8.10). Now make the value of the resistor current-dependent. Is the sum of stored energy and total dissipated energy still constant?

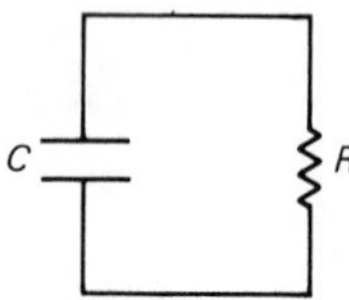

Figure 8.10 An initially charged capacitor C discharged through a resistor R.

2. Write and run a program to find the current and voltages in a series LR circuit (with no battery or generator). See Figure 8.11. Find the energy stored in the magnetic field and the total dissipated energy at each time step. Show that the sum is constant, even if the resistor is current-dependent.

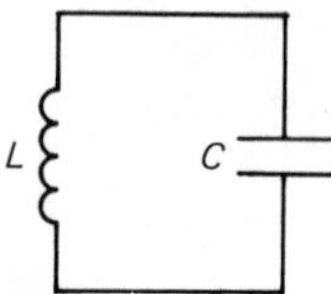

Figure 8.11 A series circuit containing a self-inductance L and a capacitor C.

3. Two circuits are coupled by a mutual inductance, M (Figure 8.12). The first circuit contains an initially charged capacitor C as well as the self-inductance L and resistance R_p of the primary coil of M. The second circuit contains the self-inductance L_2 and resistance R_2 of the second coil of M as well as a resistor R_0.
   a. Write down the (coupled) differential equations of the two circuits. Remember that mutual inductors work both ways, so both equations contain terms in M.

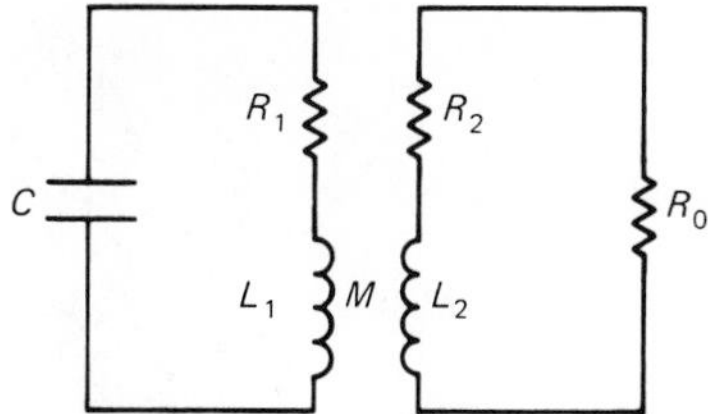

Figure 8.12 Two circuits, a series *LRC* and a series *LR*, coupled by a mutual inductor, *M*.

b. Write an algorithmic procedure to solve these equations. Try to think about the accuracy of the numerical procedure. Use $L_1 = L_2 = M = 1$ H, $C = 1$ F, $R_1 = R_2 = R_0 = 1\ \Omega$. (Although 1H coils exist, they are gigantic, and a 1F capacitor is truly immense; but 1 is a convenient number to work with.)

c. Write and run a program from your procedure. Give a physical explanation of the results that you find.

4. Draw a schematic diagram for a parallel *LRC* arrangement driven by a sinusoidal generator. Write down the differential equation(s) governing the currents and voltages in the circuit. Write a procedure to solve the equation(s) and a program embodying the procedure. Use $L_s = 10^{-2}$ H, $R = 10$ kΩ, and $C = 10^{-10}$ F. Use $E_0 = 1$ V and $\omega = 10^6$ rad/s. Then try other frequencies.

5. You can approximate the behavior of an active element (like a transistor) as a box whose output to one circuit is proportional to the current in another circuit. Consider a circuit in which the box marked $+A$ is the active element (Figure 8.13). Assume $I_2 = +AI_1$ and write down the equations for the two circuits. The circuits are coupled by the active element even though active elements often work essentially in only one direction. Write a procedure to solve the equations and write a program from your procedure. Run the program with $R_1 = 10^4\ \Omega$, $E_0 = 1$ V, $A = 50$, $R_2 = 10^4\ \Omega$, and $C_2 = 10^{-6}$ F. Use several frequencies between $\omega = 2\pi$ and $\omega = 2000\pi$. Find the phase shifts and amplitudes as functions of frequency.

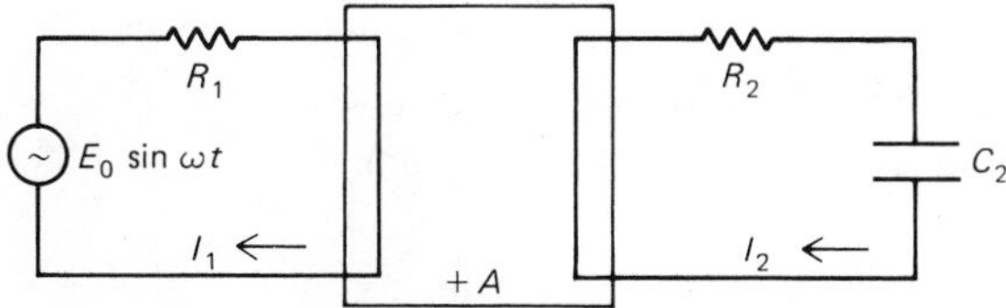

Figure 8.13 Two circuits coupled by a linear active element having amplification $+A$.

9 geometrical optics

9.1 Introduction

Geometrical optics deals directly with everyday experience. Geometrical optics deals with lenses and mirrors and is based on tracing rays of light through various media. It also deals with contact lenses, automobile headlights, mirages, the shine and glitter of jewelry, and the strange-looking world from underwater.

This chapter starts with a brief discussion of the use of the computer to solve thin-lens and simple mirror problems. Practice in the use of these simple relationships introduces you to the words of geometrical optics. You will have a better intuitive feeling for virtual images and virtual objects; you will see more pictures of object-image pairs in various simple optical systems.

The chapter continues with the simplest form of ray tracing. Instead of idealizing nature and dealing only with the paraxial rays (rays very near the optical axis) and thin lenses, you can now deal with thick, multiple lenses and rays far off axis. The strategy uses Snell's law at each interface between media. Another ray-tracing program bounces rays off mirrors. These are completely general approaches. You need only be able to treat the propagation of light in terms of rays.

The chapter ends with a discussion of situations in which the index of refraction is a continuous function of position. These problems include mirages on a hot road, loomings in the air (including sightings of the mythical Flying Dutchman), and the way radio waves go around the world.

9.2 Basic Formulas

The following facts and formulas are derived in your course text. If something looks new, review your textbook.

1. The incident, reflected, and refracted rays lie in a plane (which also includes the normal to the interface surface at the point of intersection).
2. Reflection: angle of incidence θ_1 = angle of reflection θ_2
3. Refraction: $n_1 \sin \theta_1 = n_2 \sin \theta_2$ *(Snell's law)*
4. Total internal reflection:

$$\sin \theta_{\text{critical}} = \frac{n_1}{n_2}$$

5. Polarizing angle:

$$\tan \theta_{\text{polarizing}} = \frac{n_2}{n_1}$$ *(Brewster's angle)*

6. Thin-lens, or mirror, formula:

$$\frac{1}{o} + \frac{1}{i} = \frac{1}{f}$$

where o = object distance
i = image distance
f = focal length
The sign convention is important.

7. Thin lens:

$$\frac{1}{f} = (n - 1)\left(\frac{1}{r_1} - \frac{1}{r_2}\right)$$ *(Lens-maker's formula)*

8. Spherical mirror:

$$f = \frac{r}{2}$$

Program 9.1 (a)

LENMIRBA *Thin lens and mirror calculations*

```
100 PRINT "MIRROR OR LENS SYSTEM";
110 INPUT M$
120 IF M$="MIRROR" THEN 240
130 PRINT "INPUT FOCAL LENGTH (RATHER THAN RADII,ETC.) (YES OR NO)";
140 INPUT Y$
150 IF Y$="YES" THEN 210
160 PRINT "INDEX OF REFR., RAD. OF 1ST SURF., & RAD. OF 2ND SURF.";
170 INPUT N,R1,R2
180 LET F=1/((N-1)*(1/R1-1/R2))    Focal length from lens maker's formula
190 PRINT "  FOCAL LENGTH =";F;" INCHES."
200 GOTO 230
210 PRINT "FOCAL LENGTH OF LENS";
220 INPUT F
230 GOTO 280
240 PRINT "RADIUS OF CURVATURE";
250 INPUT R1
260 LET F=R1/2    Focal length of mirror
270 PRINT "  FOCAL LENGTH =";F;" INCHES."
280 PRINT "OBJECT POSITION AND HEIGHT (999,999 TO END)";
290 INPUT X5,Y5
300 IF (X5-999)*(Y5-999)=0 THEN 410
310 IF X5<>F THEN 340
320 PRINT "  CAN'T CALCULATE INFINITE CASES."
330 GOTO 280
340 LET X6=1/(1/F-1/X5)    Thin lens or mirror formula
350 LET M=-X6/X5    Magnification
360 LET Y6=Y5*M
370 PRINT "IMAGE POSITION AND HEIGHT: ";X6;",";Y6;"INCHES."
380 PRINT "  MAGNIFICATION:";M
390 PRINT
400 GOTO 280
410 END
```

Program 9.1 (b)

LENMIRFO *Thin lens and mirror calculations*

```
         DATA ANLEN,R1LEN,R2LEN/1.5,10.,-10./    Index of refraction, radii of surfaces
         DATA RMIR/10./    Radius of mirror
         FLEN=1./((ANLEN-1.)*(1./R1LEN-1./R2LEN))    Lens maker's formula
         FMIR=RMIR/2.
5        READ(5,100) XOBJ,YOBJ    Position and height of object
         IF (XOBJ-999.) 10,50,10
10       PRINT 101,XOBJ,YOBJ
         XLIM=1./(1./FLEN-1./XOBJ)    Thin lens formula
         ALMAG=XLIM/XOBJ    Magnification
         YLIM=YOBJ*ALMAG
         XMIM=1./(1./FMIR-1./XOBJ)    Mirror formula
         AMMAG=XMIM/XOBJ    Magnification
         YMIM=YOBJ*AMMAG
         PRINT 102,FMIR,XMIM,YMIM,AMMAG
         PRINT 102,FLEN,XLIM,YLIM,ALMAG
         GOTO 5
50       STOP
100      FORMAT(2F10.4)
101      FORMAT(1X,2E12.4)
102      FORMAT(1X,4E12.4)
         END
```

9.3 Thin Lenses and Spherical Mirrors

Introductory geometrical optics derives the thin-lens, or mirror, formula. Experience with the formula teaches you the terminology of image formation.

The program LENMIR [Programs 9.1(a), (b)] solves the simple lens formula. The program asks for focal-length information and then an object position and height; the program returns the coordinates of the image (position and height). The problems at the end of the chapter include examples of the use of LENMIR.

9.4 Ray Tracing in Lens Systems

The program THLEN [Programs 9.2(a), (b), (c)] traces a ray through a series of media of different indices of refraction. The media are separated by spherical interfaces whose centers lie on the x axis. The program is simple. You supply the starting point and the angle which the initial ray makes with the x axis; the program calculates the intersection of the ray and the first interface. The angle between the normal to the surface and the incident ray is calculated and then Snell's law is used to find the refracted angle in the next medium; the angle the refracted ray makes with the x axis is calculated and the process is repeated. The method is entirely general. The program restricts the problem to spherical interfaces which occur in a definite order, but this restriction can be removed. You can write a version for general interfaces if you wish.

Program 9.2 (a) Flow chart for a thick-lens ray-tracing strategy. After setting the initial values of parameters, the calculation determines the path of a ray until a change in index of refraction occurs. Snell's law determines the new direction of the ray, and the process continues.

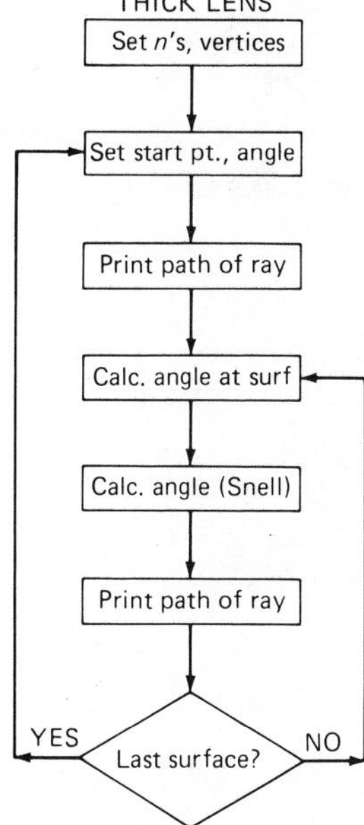

Program 9.2 (b)

THLENBAS *Ray tracing through thick lenses*

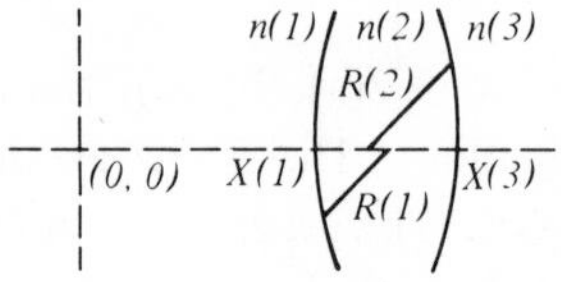

```
100 PRINT "HOW MANY DIFFERENT REGIONS";
110 INPUT N0
120 FOR I=1 TO N0-1
130 PRINT "INDEX OF REFRACTION OF REGION";I;
140 INPUT N(I)
150 PRINT "VERTEX POS. & RADIUS OF SURFACE BETWEEN REGIONS";I;"&";I+1;
160 INPUT X(I),R(I)
170 NEXT I
180 PRINT "INDEX OF REFRACTION OF LAST REGION";
190 INPUT N(I+1)
200 LET I=0
210 PRINT "Y OF SOURCE, ANGLE TO AXIS OF RAY (DEGS.)";
220 INPUT Y0,A
230 LET A=3.14159*A/180
240 LET X1=0
250 LET Y1=Y0          Initial point on ray
260 PRINT X1,Y1
270 LET I=I+1
280 IF I=N0 THEN 650
290 LET T=TAN(A)       Slope of ray
300 LET R=R(I)
310 LET X7=X(I)+R
320 LET E=T*T+1
330 LET F=2*(Y0*T-X7-T*T*X1)
340 LET G=Y0*Y0+X7*X7-R*R+T*T*X1*X1-2*Y0*T*X1
350 LET F1=F*F-4*E*G
      (320-350: Coefficients in quadratic equation for intersection of ray and next surface)
360 IF F1>0 THEN 400
370 PRINT "  MISSED NEXT SURFACE."
380 PRINT
390 GOTO 200
400 IF R<0 THEN 430
410 LET X0=(-F-SQR(F1))/(2*E)
420 GOTO 440
430 LET X0=(-F+SQR(F1))/(2*E)
      (400-430: x of intersection of ray with surface)
440 LET Y0=Y0+T*(X0-X1)      y of intersection
450 LET X1=X0
460 LET Y1=Y0
470 PRINT X1,Y1        Next point on ray
480 LET A1=ATN(Y0/(X7-X0))   Angle of ray to x axis
490 LET A3=A1
500 LET A0=A+A3        Angle of ray to surface normal
510 FOR J=1 TO 4
520 IF ABS(A0)-J*3.14159/2>0 THEN 550     Find quadrant
530 LET J0=J-1
540 GOTO 560
550 NEXT J
560 LET S1=SIN(A0)
570 LET S2=N(I)*S1/N(I+1)    Snell's law at surface
580 IF ABS(S2)<1 THEN 610
590 PRINT "  TOTAL INTERNAL REFLECTION."
600 GOTO 200
610 LET A2=ATN(S2/SQR(1-S2*S2))
620 LET A2=A2-SGN(A0)*J0*3.14159/2
630 LET A=A2-A3
      (610-630: Find new angle)
640 GOTO 270
650 LET Y1=Y1+TAN(A)*(10-X1)
660 LET X1=10
670 PRINT X1,Y1
      (650-670: Final point on ray)
680 GOTO 200
690 END
```

n(1) n(2) n(3)
R(2)
(0, 0) X(1) X(3)
R(1)

Program 9.2 (c)

THLENFOR *Ray tracing through thick lenses*

```
      DIMENSION AN(3),VERT(2),RAD(2)
      DATA AN(1),VERT(1),RAD(1)/1.,-5.,3./             } Indices of refraction,
      DATA AN(2),VERT(2),RAD(2),AN(3)/2.,5.,-3.,1./    } vertices of surfaces,
                                                       } radii of surfaces
10    READ(5,101) Y,ANG     Initial point and angle of ray
      IF (Y.EQ.9999.) GOTO 70
      ANG=3.14159*ANG/180.
      X=0.
      PRINT 104,X,Y     First point on ray
      TANANG=SIN(ANG)/COS(ANG)
      DO 60 I=2,3,1
      XCENT=VERT(I-1)+RAD(I-1)
      ASQR=TANANG**2+1.                                              } Coefficients
      BSQR=2.*(Y*TANANG-XCENT-X*TANANG**2)                            } of quadratic
      CSQR=Y**2+XCENT**2-RAD(I-1)**2+(TANANG*X)**2-2.*Y*TANANG*X      } formula for
      DSQR=BSQR**2-4.*ASQR*CSQR                                       } intersection
      IF (DSQR) 20,30,30                                              } of ray and
20    PRINT 102                                                       } surface
      GOTO 10
30    XTEMP=X
      X=(-BSQR-(RAD(I-1)/ABS(RAD(I-1)))*SQRT(DSQR))/(2.*ASQR)  } Point on ray at next
      Y=Y+TANANG*(X-XTEMP)                                     } surface
      PRINT 104,X,Y     Next point on ray
      ANGTOX=ATAN(Y/ABS(XCENT-X))     New angle of ray to x axis
      ANGTON=ANGTOX+(RAD(I-1)/ABS(RAD(I-1)))*ANG     Angle to surface normal
      SINNEW=AN(I-1)*SIN(ANGTON)/AN(I)     Snell's law
      IF (ABS(SINNEW)-1.) 50,50,40   } Total internal
40    PRINT 103                      } reflection
      GOTO 10
50    ANG=ATAN(SINNEW/SQRT(1.-SINNEW**2))                 } New angle
      ANG=-(RAD(I-1)/ABS(RAD(I-1)))*(ANGTOX-ANG)          } of ray
60    TANANG=SIN(ANG)/COS(ANG)     New slope of ray
      Y=Y+TANANG*(10.-X)   } Final
      X=10.                } point on
      PRINT 104,X,Y        } ray
      GOTO 10
101   FORMAT(2F5.2)
102   FORMAT(1X,20HMISSED LENS SURFACE.)
103   FORMAT(1X,26HTOTAL INTERNAL REFLECTION.)
104   FORMAT(1X,2E15.5)
70    STOP
      END
```

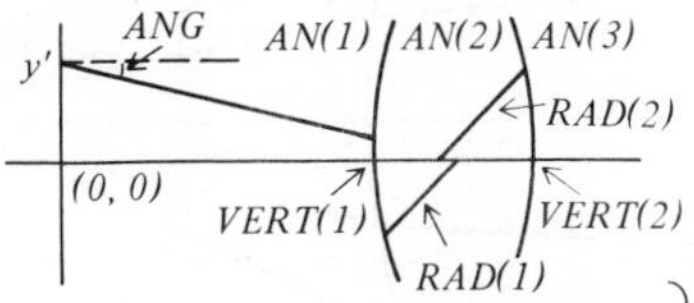

You now have a diagrammatic way to visualize various aberrations of a lens system. For example, spherical aberration is easily observed in plots of the output of the program. Figure 9.1 is a plot of parallel rays passing through a double convex lens. For paraxial rays, the point at which these rays cross the axis is unique (the focal point), but you can see that the rays far off axis focus to different points. Some rays meet the second surface at angles so large that the rays are totally internally reflected. Figure 9.2 is a plot of parallel rays passing through a double concave lens. If the lens were thin and the rays paraxial, the final rays would appear to diverge from a single point, the focus. For thick lenses, the rays extrapolate back to different places. Figure 9.3 shows an interesting effect. The lens is a bent sheet of plane-parallel glass. According to thin-lens theory, the focal length of such a lens is infinite; rays can be displaced,

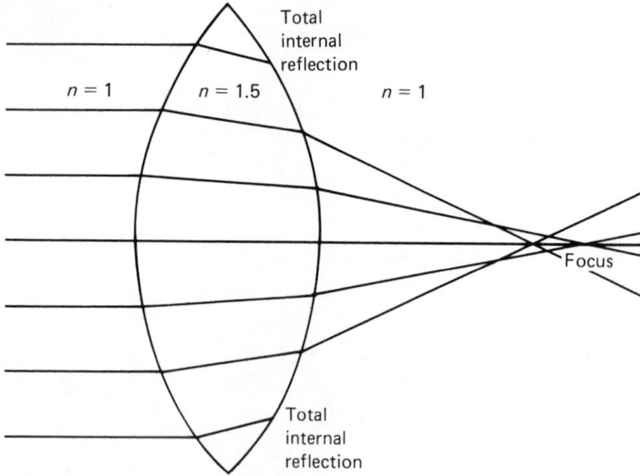

Figure 9.1 Ray tracing through a thick lens. Using Snell's law at each interface, rays are traced through a lens system. The paraxial rays through the double convex lens meet at the focal point. Rays further off axis demonstrate spherical aberration. Rays far off axis are totally internally reflected.

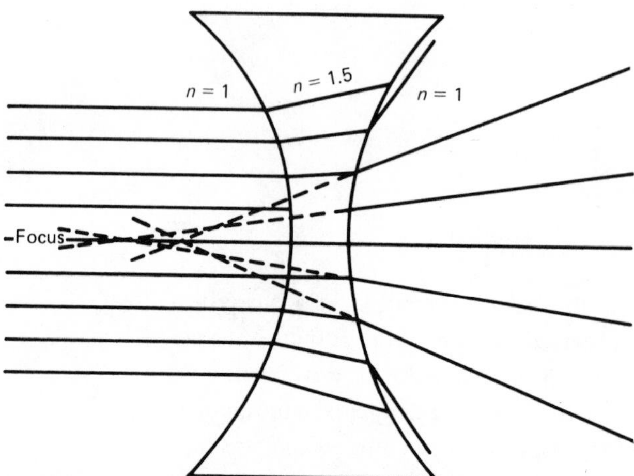

Figure 9.2 Ray tracing through a thick diverging lens. Snell's law is used at each interface to trace rays through the lens system. Paraxial rays appear to diverge from a common focal point. Rays further off axis demonstrate spherical aberration. Rays far off axis undergo total internal reflection.

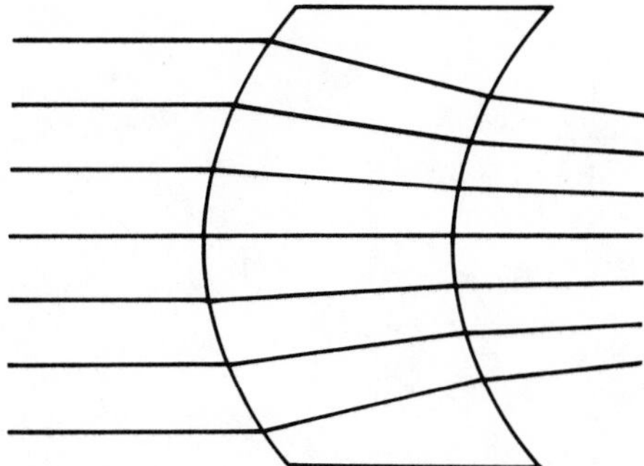

Figure 9.3 Rays traced through a bent sheet. Paraxial rays remain parallel. Rays far off show a weak focusing effect.

but they must remain parallel. However, using ray-tracing methods, you see that there is some focusing effect from this lens.

In all these cases, the rays, which are close to the axis and make only small angles to the axis (paraxial rays), do obey thin-lens theory. Using the ray-tracing technique allows you to deal with a wider range of situations; but under the approximations of thin-lens theory, the simpler theory holds.

Similar calculations using ray-tracing methods for rays reflected from mirror surfaces can be performed [MIRROR, Programs 9.3(a), (b)].

Program 9.3 (a)

```
MIRRORBA    Rays off mirror

100 PRINT "RADIUS, RADIUS OF CURVATURE";
110 INPUT R,R1
120 PRINT "DISTANCE OF SOURCE FROM MIRROR, HEIGHT ABOVE AXIS";
130 INPUT L,L1
140 PRINT "# OF STEPS FROM ";-R;"TO";R;
150 INPUT S
160 FOR I=-R TO R STEP 2*R/S     Step up mirror
170 LET X1=L        } Initial point
180 LET Y1=L1       } on ray
190 PRINT X1,Y1     }
200 LET Y1=I                                        } Ray at
210 LET X0=R1-SQR(R1*R1-I*I)    Spherical mirror   } mirror
220 LET X1=X0                                       }
230 PRINT X1,Y1                                     }
240 LET T1=ATN(I/(R1-X0))
250 LET T2=ATN((I-L1)/(L-X0))
260 LET T3=-(2*T1-T2)
270 LET T4=TAN(T3)    Slope of ray off mirror
280 LET X1=10              } Point on
290 LET Y1=I+T4*(X1-X0)    } ray leaving
300 PRINT X1,Y1            } mirror
310 PRINT
320 NEXT I
330 END
```

Program 9.3 (b)

MIRRORFO *Rays off mirror*

```
          DATA RAD,RADCU/3.,5./     Radius, radius of curvature
          DATA XSOUR,YSOUR,NSTEP/10.,1.,8/    Source point, number of rays
          DO 60 I=1,NSTEP+1     Step up mirror
          YMIR=-RAD+(I-1)*2.*RAD/NSTEP
          X=XSOUR    }
          Y=YSOUR    } Initial point on ray
          PRINT 100,X,Y }
          XMIR=RADCU-SQRT(RADCU**2-YMIR**2)    Spherical mirror  }
          X=XMIR                                                  } Ray at mirror
          PRINT 100,X,YMIR                                        }
          ATN1=ATAN2(YMIR,RADCU-X)
          ATN2=ATAN2(YMIR-YSOUR,XSOUR-X)
          ATN3=-(2.*ATN1-ATN2)
          TN3=SIN(ATN3)/COS(ATN3)    Slope of ray off mirror
          X=10.                  }
          Y=YMIR+TN3*(X-XMIR)    } Point on ray leaving mirror
60        PRINT 100,X,Y          }
          STOP
100       FORMAT(1X,2F10.4)
          END
```

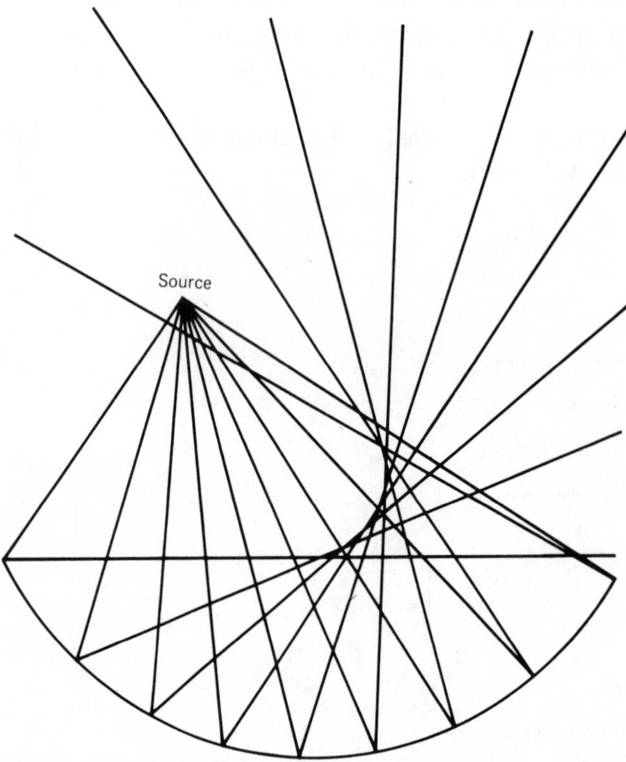

Figure 9.4 Ray tracing for a spherical mirror. The equal-angle formula is used for the reflections of eight points on the mirror. Spherical aberration is shown.

Figure 9.4 shows rays reflected from a mirror. Again, you can observe spherical aberration effects. You can also discover ring-of-light effects for rays from a point source of light just off axis near the focal point of a paraboloidal mirror. For both lenses and mirrors, you can show that surfaces can be shaped so that a single image point occurs for all rays from a single object point. Then you can discover how hard it is to shape lens surfaces or mirror surfaces so that all the points on an extended source are imaged perfectly onto a single replica. These are some of the problems of optical system design; much more sophisticated ray-tracing methods have been developed to handle them.

9.5 Ray Tracing in Continuous Media

Ray-tracing methods can be extended to situations in which indices of refraction are continuous functions of position. Such situations occur in mirages and related phenomena. When the ground is hotter than the air, a sheet of air having an index of refraction close to 1 lies just above the ground. Light rays are bent as they enter this rarified region. Some of them are totally internally reflected back into the upper air. The rays produce a virtual image of the object they left; the image is inverted and appears below the ground. Typical road mirages are images of the sky.

MRAG is a program which follows rays through media where the index of refraction n is a function of the height y, $n = n(y)$. MRAG [Programs 9.4(a), (b), (c)] uses Snell's law at each point on the ray. You follow the ray step by step. Consider moving a

Program 9.4 (a) Block diagram of the strategy to calculate rays through a mirage. The index of refraction varies with the height y above the ground.

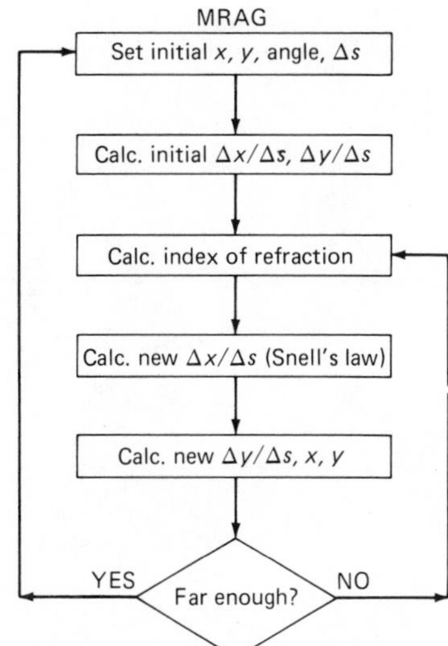

Program 9.4 (b)

MRAGBAS *Mirage ray tracing*

```
100 PRINT "INITIAL (X,Y) AND ANGLE OF RAY TO X-AXIS (DEGS.)";
110 INPUT X1,Y1,T
120 LET T=90-T                    } Angle to vertical
130 LET T=3.14159*T/180           } in radians
140 LET I=0
150 LET D3=SIN(T)     Initial Δx/Δs
160 LET D2=COS(T)     Initial Δy/Δs
170 LET S9=SGN(D2)
180 PRINT "X","Y"
190 LET I=I+1
200 LET D1=.1       Step size, Δs
210 LET N0=N
220 LET Y=Y1+D2*D1/2    Half-stepped height
230 LET N=2             }
240 IF Y>3 THEN 260     } Index of
250 LET N=1+.1*Y*Y      } refraction
260 IF I>1 THEN 290     }
270 LET N0=N            } Initial
280 LET D1=0            } index of refraction
290 LET D3=N0*D3/N    Snell's law; new Δx/Δs
300 IF ABS(D3)<=1 THEN 340  }
310 LET S9=-S9              } Total internal
320 LET D2=-D2              } reflection
330 GOTO 350                }
340 LET D2=S9*SQR(1-D3*D3)    New Δy/Δs
350 LET Y1=Y1+D2*D1      New y
360 LET X1=X1+D3*D1      New x
370 IF ABS(X1-X0)+ABS(Y1-Y0)<.5 THEN 410  }
380 PRINT X1,Y1                           } Print
390 LET X0=X1                             } group
400 LET Y0=Y1                             }
410 IF X1*(X1-10)>0 THEN 430    } Test for
420 IF Y1*(Y1-10)<=0 THEN 190   } off page
430 PRINT
440 GOTO 100      Return for new ray
450 END
```

Program 9.4 (c)

```
MRAGFOR     Mirage ray tracing

10     READ(5,100)X,Y,ANG       Initial position and angle of ray
       IF (X.EQ.9999.) GOTO 80
       ANG=(90.-ANG)*3.14159/180.     Angle to vertical in radians
       PDIST=0.
       DXDS=SIN(ANG)      Initial Δx/Δs
       DYDS=COS(ANG)      Initial Δy/Δs
       SIGNDY=DYDS/ABS(DYDS)
       DS=.001      Step size, Δs
       AN=Y-DYDS*DS
20     ANOLD=AN
       AN=Y
       DXDS=ANOLD*DXDS/AN      Snell's law; new Δx/Δs
       IF (ABS(DXDS)-1.) 40,40,30  }
30     SIGNDY=-SIGNDY              } Total
       DYDS=-DYDS                  } internal
       GOTO 45                     } reflection
40     DYDS=SIGNDY*SQRT(1.-DXDS**2)      New Δy/Δs
45     Y=Y+DYDS*DS      New y
       X=X+DXDS*DS      New x
       PDIST=PDIST+DS               }
       IF (PDIST-.5) 60,60,50       } Print
50     PRINT 101,X,Y                } group
       PDIST=0.                     }
60     IF (X*(X-10.)) 70,70,10      } Test for
70     IF (Y*(Y-7.)) 20,10,10       } off page
100    FORMAT(3F5.2)
101    FORMAT(1X,2E15.5)
80     STOP
       END
```

distance ds along a ray and let $n = n(y)$. See Figure 9.5. From Snell's law, at each step

$$n_{\text{new}} \sin\theta_{\text{new}} = n_{\text{old}} \sin\theta_{\text{old}} \quad \text{or} \quad n_{\text{new}} \left(\frac{dx}{ds}\right)_{\text{new}} = n_{\text{old}} \left(\frac{dx}{ds}\right)_{\text{old}}$$

so

$$dx = ds \frac{n_{\text{old}}}{n_{\text{new}}} \left(\frac{dx}{ds}\right)_{\text{old}}$$

$$dy = \sqrt{(ds)^2 - (dx)^2}$$

You then get the new position $(x + dx, y + dy)$. The method is very reminiscent of $\mathbf{F} = ma$ programs or field-line plotting. Since you do not know n_{new}, you approximate it using information you already know. For example, you can guess that the change in n at this step is going to be about the same as at the last step. You can also use half-step approaches to increase accuracy.

Figure 9.6 is a plot of some rays from an object. The index of refraction for the figure is a model of the variation of n over heated ground. Some rays reach an observer

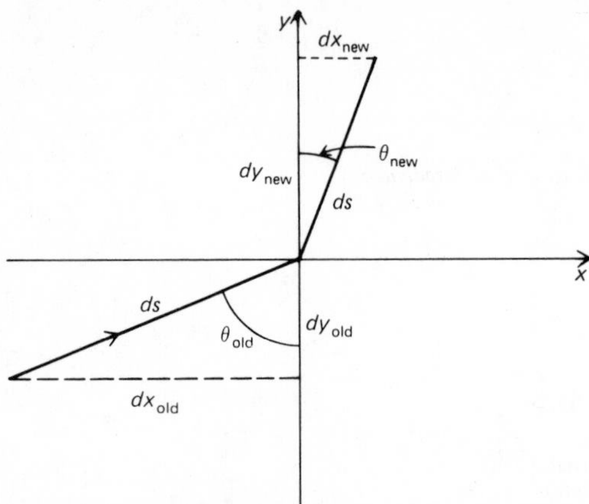

Figure 9.5 The geometry of a Snell's law calculation for mirage and looming calculations. The light ray comes from the old side of the interface and goes to the new side.

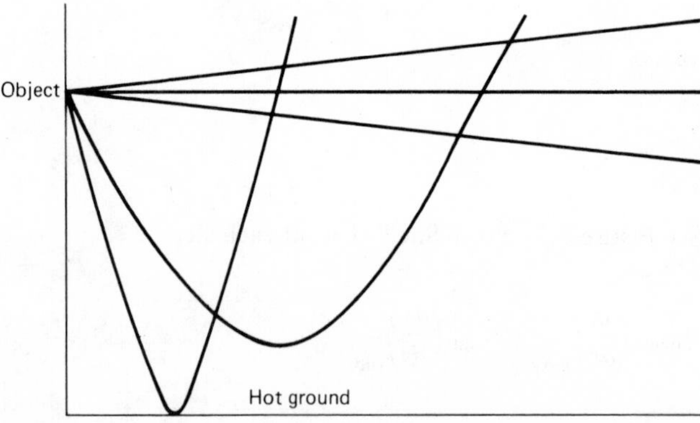

Figure 9.6 Ray tracing through a mirage region. The index of refraction varies with height above the heated ground. Rays are bent so that an observer sees the object and an image of the object. The image is inverted and below ground; the image is the mirage.

at the right of the figure directly from the object; other rays have been bent and appear to originate from a point below the ground. You will find that although qualitative features of the rays do not depend on the exact way n varies with height, the quantitative prediction of where rays will travel and where mirage images appear depends quite strongly on your model of $n(y)$. There is interesting physics in these problems, including some very nice work in the propagation of radio waves through the ionosphere.

9.6 Waves Off Mirrors

The earlier sections of this geometrical optics chapter described phenomena in terms of rays. For objects large with respect to the wavelength of the wave, the description is perfectly adequate. Sometimes it is easy to lose sight of wave patterns when dealing only with rays.

One way to see wave patterns off obstacles is by means of the series of water wave film loops which show wave reflections. A second way uses the computer to follow waves bouncing off obstacles. The computer method is demonstrated by the program SPHR [Programs 9.5(a), (b)]. The wave pulse starts at a source S and propagates at speed v for some time t. At time t the computer checks points across the mirror to see

Program 9.5 (a)

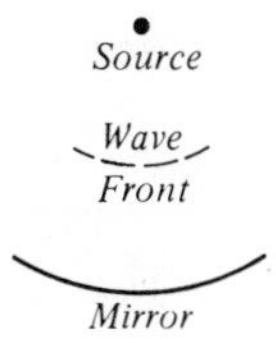

```
SPHRBAS      Spherical wavefront off mirror

100 PRINT "SOURCE OF PULSE AND ITS VELOCITY";
110 INPUT S1,S2,V
120 PRINT "TIME";
130 INPUT T           Time front has propagated
140 LET D=T*V         Distance front has propagated
150 FOR J=-5 TO 5 STEP .5       Step across x
160 LET F=J*J/4         Shape of mirror
170 LET F1=J/2        First derivative of shape
180 LET F2=SQR(1+F1*F1)
190 LET Y2=S2-F                 }
200 LET X2=S1-J                 } r from source
210 LET R=SQR(X2*X2+Y2*Y2)      } to point on mirror
220 IF D<R THEN 330       Has not reached mirror
230 LET D0=D-R        Distance after bounce
240 LET N1=-F1/F2     } Unit normal
250 LET N2=1/F2       } vector
260 LET C=X2/R*N1+Y2/R*N2   } sin and cos of angle
270 LET S=Y2/R*N1-X2/R*N2   } between radius and normal
280 LET D1=C*N1+S*N2     } Unit vector of
290 LET D2=C*N2-S*N1     } bounced front
300 LET X=J+D1*D0        }
310 LET Y=F+D2*D0        } Point on bounced front
320 GOTO 350
330 LET X=S1-X2/R*D      }
340 LET Y=S2-Y2/R*D      } Point on unbounced front
350 PRINT X,Y
360 NEXT J
370 GOTO 120      Return for new propagation time
380 END
```

Source

Wave Front

Mirror

Program 9.5 (b)

```
SPHRFOR   Spherical wavefront off mirror

          DATA XSOUR,YSOUR,V/2.,0.,1./     Source, velocity
10        READ(5,100)T     Propagation time
          IF (T-999.) 20,90,90
20        D=T*V     Distance propagated
          DO 80 J=1,21     Step across x
          AJ=-5.+(J-1)*.5
          F=AJ*AJ/4.     Shape of mirror
          F1=AJ/2.     First derivative of shape
          F2=SQRT(1.+F1*F1)
          Y2=YSOUR-F                  } r from source
          X2=XSOUR-AJ                 } to point on mirror
          R=SQRT(X2*X2+Y2*Y2)
          IF (D-R) 70,30,30     Has not reached mirror
30        D0=D-R     Distance after bounce
          AN1=-F1/F2     } Unit normal
          AN2=1./F2      } vector
          C=X2/R*AN1+Y2/R*AN2   } sin and cos of angle
          S=Y2/R*AN1-X2/R*AN2   } between radius and normal
          D1=C*AN1+S*AN2   } Unit vector
          D2=C*AN2-S*AN1   } along bounced direction
          X=AJ+D1*D0   } Point on bounced front
          Y=F+D2*D0    }
          GOTO 80
70        X=XSOUR-X2/R*D   } Point on unbounced front
          Y=YSOUR-Y2/R*D   }
80        PRINT 101,X,Y
          GOTO 10     Return for new propagation time
90        STOP
100       FORMAT(F10.5)
101       FORMAT(1X,2F10.4)
          END
```

Source

Wave Front

Mirror

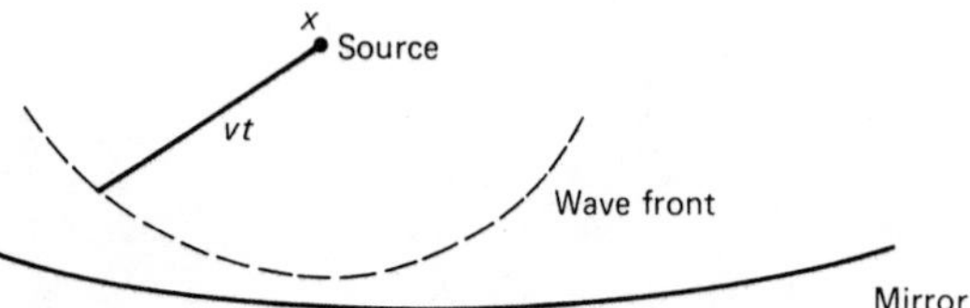

Figure 9.7 The geometry of a wavefront emitted from a source and about to reach a mirror. The wavefront has traveled for a time t at speed v.

if the part of the pulse heading toward the point has reached the mirror. The present point on the wavefront is on the spherical wavefront a distance $D = vt$ from the source (see Figure 9.7). If the wavefront has already reached the point on the mirror, the computer calculates where that piece of the wave ended up at time t. The calculation involves finding the distance D_0 the wavefront traveled after hitting the mirror. The calculation also uses the sine and cosine of the angle A between the direction the wave

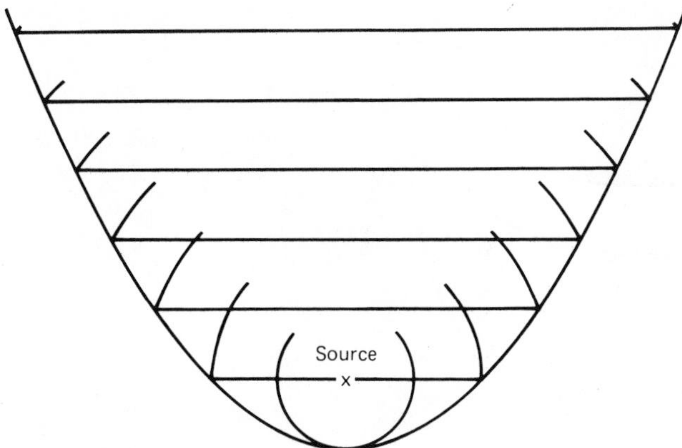

Figure 9.8 Spherical waves reflecting from the focal point of a paraboloidal mirror. The final wavefront is plane parallel. The source is marked as an **x**. The wavefront is shown after seven equal time intervals.

is moving and the normal to the mirror. This angle A is also the angle the final wave direction makes with the normal; the computer propagates the wavefront a distance D_0 in the direction A from the mirror.

The procedure is general. The user defines the shape of the mirror and supplies an equation for the first derivative. The mirror shape must be single-valued in x. The routine SPHR uses the dot product of vectors to find the cosine of the angle A and the cross product to find the sine of A. This use of vector products is very fast on the computer because only arithmetic operations (+, −, *, and /) are used.

Figure 9.8 shows spherical waves emitted from the focal point of a paraboloidal mirror. Wavefronts propagated for several different times (in equal time steps) are shown; the source is marked by an **x**. All the wavefronts are parallel after leaving the mirror. The final pulse is plane-parallel. The opposite effect, that plane-parallel light is focused onto the focal point, is used in high-quality optical telescopes.

Figure 9.9 shows spherical waves reflected from a spherical mirror. The wave source is at the focal point, half the radius from the vertex of the mirror. The final waves are very nearly parallel near the center of the pattern, but the edges display spherical aberration. The wavefronts are numbered in inverse order of time; the source is marked by **x**.

Figure 9.10 shows waves and an ellipsoidal mirror. The source S lies at the focal point of the parabola having the same curvature as the ellipse at the vertex. The final waves are essentially parallel near the center but show considerable (ellipsoidal) aberration near the edges.

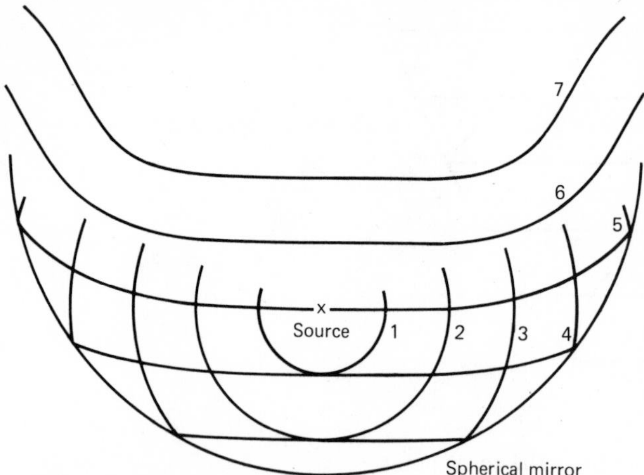

Figure 9.9 Spherical waves reflecting from the focal point of a spherical mirror. The source is shown as an **x** and the wavefronts are numbered. The final wavefront is plane-parallel near the center but shows spherical aberration at the edges.

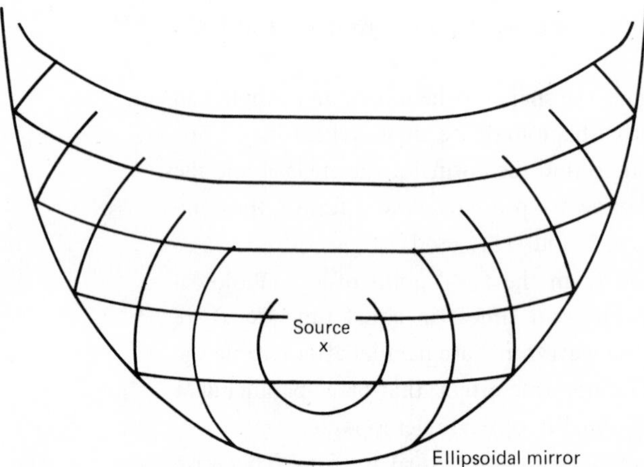

Figure 9.10 Waves reflecting from an ellipsoidal mirror. The source is at the focal point of the equivalent parabola, the parabola having the same curvature at the vertex. The final waves are nearly parallel but show aberration.

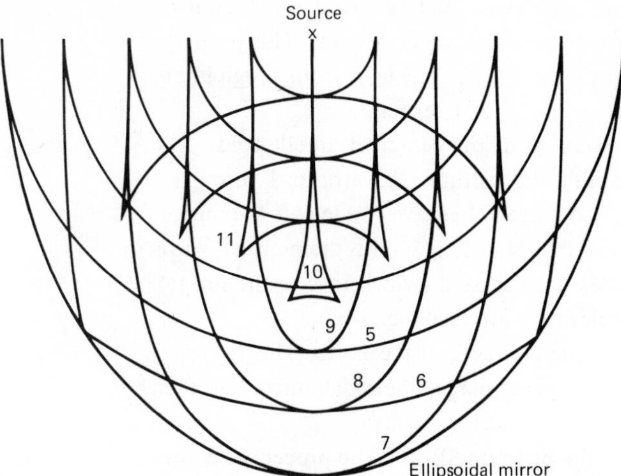

Figure 9.11 Waves reflecting from the center of an ellipsoid. Several wavefronts are numbered. After reflecting from the mirror, the wavefront crosses itself and forms an inverted spherical final wave.

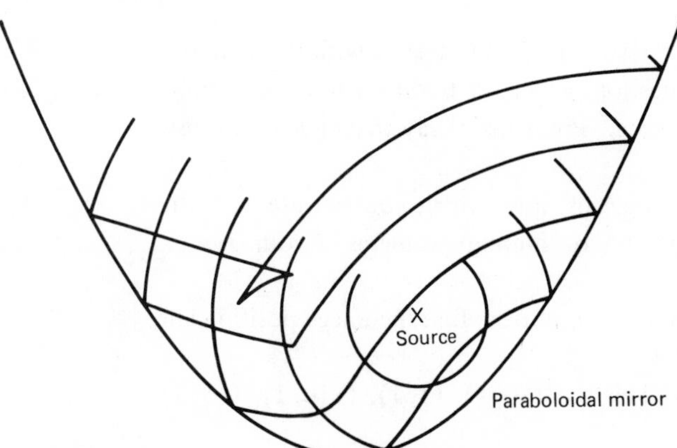

Figure 9.12 Waves reflecting from a paraboloidal mirror with the source off axis. The wavefront crosses itself after reflection and forms an inverted image.

With ellipsoidal mirrors the program can also show that spherical waves emitted from one focal point of the ellipse are focused onto the second focus. This property of elliptical mirror arrangements is used in pulsed lasers. The light from a high-intensity flash tube is focused onto the laser rod to trigger the laser flash.

Figure 9.11 shows waves propagating away from the center of an ellipsoid. The wavefronts are numbered in order to simplify the picture. Wavefronts 1 through 5 have not reached the mirror. Wavefronts 6 and 7 show the wave coming off the mirror. The curves numbered 8 and 9 show the wavefronts closing in onto themselves. Wavefront 10 shows the wave just after the sides have crossed (which inverts the image). Wavefronts 11 through 13 show the final spherical wave forming.

Figure 9.12 illustrates the effects when the source is off center. The source is marked by a **x**. The spherical wavefront bounces off the paraboloidal mirror and then starts to invert. The final wave will be nearly spherical.

These figures are plotted from output of the program SPHR. The procedure is very general and works for any type of wave off any mirrorlike obstacle. A very similar technique can be used for refracted waves, too. Even combinations of transmission and reflection can be treated.

Problems

As always, there are a large number of interesting problems dealing with this material. Some of the problems are extensions of questions which are traditional. In these problems, the computer helps you gain experience. Problems of ray tracing have not usually been given in traditional courses.

The following problems are a small sample of those which you can use with this material. The problems are only prototypes; try to think up problems of your own.

1. Given the following information about a mirror, find the image position and height for each object:
   a. $r = 10$ cm; objects (position, height): (10, 1), (5, 1), (-10, 1), (-5, 1), (10, -1)
   b. $r = -10$ cm; objects as in part a
2. Given the following information about a lens, find the image position and height for each object:
   a. $n = 1.5$, $r_1 = 10$, $r_2 = -10$; objects (positions, height): (10, 1), (-10, 1), (2.5, 1), (-2.5, 1)
   b. $n = 1.5, r_1 = -10, r_2 = 10$; objects as in part a
3. a. Write a program to find the angle of reflection of any ray off any point on a spherical mirror. Remember that it is the angle between the ray and the normal that counts.

b. Using your program, find and plot the trajectories of a number of rays from some object. Demonstrate aberration.

c. Demonstrate that paraxial rays obey the simple theory given in your course text.

4. a. Write a program that traces rays through a lens having two spherical interfaces with air but a choosable index of refraction and surface radii.

   b. Execute your program. Find and plot the trajectories of a number of rays. Demonstrate aberration.

   c. Demonstrate, using your program, that a thin-lens-paraxial-ray theory holds for appropriate lenses and rays.

   d. Indices of refraction are normally functions of wavelength (color). Demonstrate color aberration by using your program with the same rays but two different indices of refraction—one for red light and one for blue.

5. A model of the variation of n above heated ground has

   $$n = 1 + n_0 \; [1 - \exp(-y/y_0)]$$

   where $n_0 + 1 = n$ far above the surface and y_0 = a characteristic height.

   a. Write a program which traces rays through a medium in which $n = n(y)$.

   b. Using the model of n, find the trajectories of a number of rays leaving some point source at a height y. Trace rays going up and going down from the object.

10 fourier analysis and wave propagation

10.1 Introduction

Any periodic function can be represented by a Fourier series; and almost any function at all can be represented by a Fourier transform. An understanding of Fourier analysis paves the way for a more complete understanding of the subject of wave propagation, one of the most important areas of physics. Waves occur throughout nature—not only as sound or water waves but also as matter waves (quantum mechanical particles). Many of the features which will be discussed in these next two chapters on waves are features found over and over again in physics.

This chapter first introduces you to programs which will construct Fourier series for any periodic waveform. With this groundwork, the chapter continues with a discussion of programs on traveling wave pulses and the uses of Fourier transforms. It is often possible to write down solutions to various physical situations in terms of functions such as sines and cosines, Bessel functions, or Legendre polynomials. The expansion of arbitrary waves into sine and cosine waves by means of Fourier transform theory is very similar to expansions in terms of other functions. Solutions to equations are often easier to interpret when you expand the solution in terms of simple functions.

The chapter explains how to expand periodic functions in Fourier series and then shows Fourier transforms, an easy step from Fourier series. Finally, you will be led through the study of pulse propagation. It will then be easy to study waves propagating through real media.

Almost all introductory physics courses are limited to situations in which all frequency components of a wave propagate together with the same velocity (nondispersive wave propagation). For some types of waves, this approximation is close enough to being valid so that the calculations are useful. However, more realistic views of nature take account of the fact that for most waves, some frequencies travel faster than others. Just as the thin-lens theory is a useful first approximation to the more general ray-tracing methods, so also is nondispersive wave theory a useful first approximation to the more general dispersive-wave theory discussed here. You will find that dispersive-wave theory is easy to handle with a computer approach.

10.2 Basic Equations

Here are a few basic facts and equations which are derived and discussed in your course textbook.

1. Nondispersive traveling waves: Any traveling wave (of any shape) propagates with phase velocity v, the rate at which any particular phase moves as time progresses.

   For a scalar, plane wave: If $f(x)$ specifies the shape of the wave at $t = 0$, then $f(x - vt)$ specifies the shape after a time t. [$f(x + vt)$ would be a wave going the opposite way. This chapter treats such a case as a negative v.]
2. Fourier theorem: Any function $f(x)$ that is sufficiently well behaved can be expanded in terms of sines and cosines as

$$f(x) = \int_{-\infty}^{\infty} [a(k) \cos (kx) + b(k) \sin (kx)]\, dk$$

where

$$a(k) = \frac{1}{2\pi} \int_{-\infty}^{\infty} f(x) \cos (kx)\, dx$$

and

$$b(k) = \frac{1}{2\pi} \int_{-\infty}^{\infty} f(x) \sin (kx)\, dx$$

[Sufficiently well behaved means that $\int_{-\infty}^{\infty} |f(x)| dx$ exists and f is differentiable.]

3. Fourier series: If the function f is periodic, the expansion of f will contain only harmonics of the fundamental frequency of the periodicity. For a function f of period L,

$$f(x) = \frac{A_0}{2} + \sum_{n=1}^{\infty} \left[A_n \cos\left(\frac{2\pi nx}{L}\right) + B_n \sin\left(\frac{2\pi nx}{L}\right)\right]$$

where

$$A_n = \frac{2}{L}\int_0^L f(x) \cos\left(\frac{2\pi nx}{L}\right) dx$$

and

$$B_n = \frac{2}{L}\int_0^L f(x) \sin\left(\frac{2\pi nx}{L}\right) dx$$

4. In any (linear) medium, each frequency of the wave propagates independently of the others. The components may or may not have equal velocities of propagation. (They also might not be absorbed equally.)

10.3 Fourier Series

Any reasonably well behaved function can be expanded in a Fourier series in terms of sines and cosines. Closed-form solutions for A_n and B_n can be found for many functions. For example, consider the function (Figure 10.1)

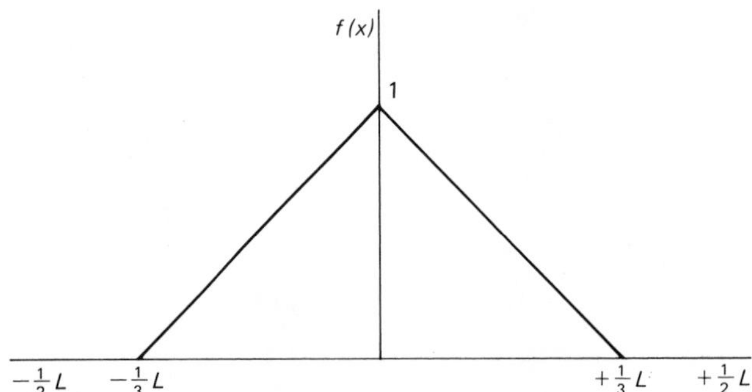

Figure 10.1 A truncated triangle wave. The wavelength or period is L.

$$f(x) = \begin{cases} 0, & \frac{-L}{2} < x < \frac{-L}{3} \\ 1 + \frac{3}{L}x, & \frac{-L}{3} < x < 0 \\ 1 - \frac{3}{L}x, & 0 < x < \frac{L}{3} \\ 0, & \frac{L}{3} < x < \frac{L}{2} \end{cases}$$

This is an even function of x, that is, $f(-x) = f(x)$; so there are no sine terms in its Fourier expansion. All the B_n are zero, and we need only calculate the A_n.

$$A_n = \frac{2}{L} \int_{-L/2}^{L/2} f(x) \cos\left(\frac{2\pi nx}{L}\right) dx$$

$$= \frac{2}{L}\left[\int_{-L/3}^{0} \left(\frac{1+3x}{L}\right) \cos\left(\frac{2\pi nx}{L}\right) dx + \int_{0}^{L/3} \left(\frac{1-3x}{L}\right) \cos\left(\frac{2\pi nx}{L}\right) dx\right]$$

$$= \frac{1}{\pi n} \sin\left(\frac{2\pi nx}{L}\right)\Bigg|_{-L/3}^{L/3} + \frac{6}{L^2}\left[\frac{L^2}{4\pi^2 n^2} \cos\left(\frac{2\pi nx}{L}\right) + \left(\frac{L}{2\pi n}\right) \sin\left(\frac{2\pi nx}{L}\right)\right]\Bigg|_{-L/3}^{0}$$

$$- \frac{6}{L^2}\left[\frac{L^2}{4\pi^2 n^2} \cos\left(\frac{2\pi nx}{L}\right) + \left(\frac{L}{2\pi n}\right) \sin\left(\frac{2\pi nx}{L}\right)\right]\Bigg|_{0}^{L/3}$$

Finally,

$$A_n = \frac{3}{\pi^2 n^2}\left[1 - \cos\left(\frac{2\pi n}{3}\right)\right]$$

$$A_0 = \frac{2}{L}\int_0^L f(x)\, dx = \frac{2}{3}$$

We can now write the Fourier series for $f(x)$:

$$f(x) = \frac{1}{3} + \sum_{n=1}^{\infty} \frac{3}{\pi^2 n^2}\left[1 - \cos\left(\frac{2\pi n}{3}\right)\right] \cos\left(\frac{2\pi nx}{L}\right)$$

Program 10.1 (a) Flow chart for a Fourier synthesis strategy. After setting the initial parameters, the strategy calculates, for each point, the sum of the Fourier components at that point.

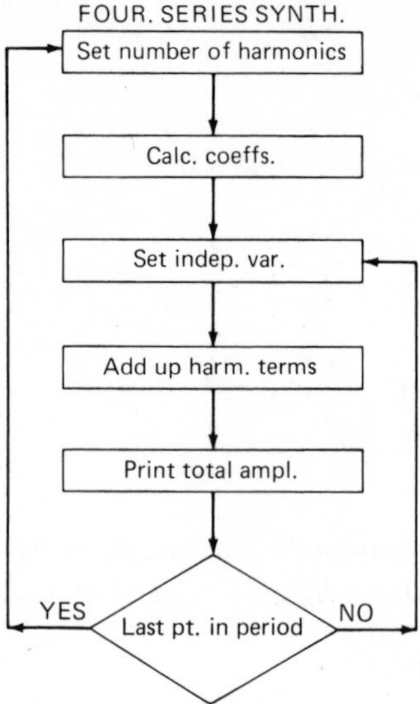

Program 10.1 (b)

SPIKEBAS *Synthesis of truncated triangle wave*

```
100 LET P=3.14159    π
110 LET L=3    Period
120 LET H=50    Number of harmonics
130 PRINT "X","AMPLITUDE"
140 FOR X=0 TO L STEP L/(5*H)    Step across x
150 LET Y=1/3
160 FOR N=1 TO H    Sum over harmonics
170 LET A=3/(P*P*N*N)*(1-COS(2*P*N/3))    Coefficient
180 LET Y=Y+A*COS(2*P*N*X/L)    Amplitude
190 NEXT N
200 PRINT X,Y
210 NEXT X
220 END
```

Program 10.1 (c)

SPIKEFOR *Synthesis of truncated triangle wave*

```
        PI=3.14159
        ALEN=3.    Period
        NHARM=50    Number of harmonics
        NSTEP=5*NHARM+1
        ANSTEP=NSTEP-1
        DO 20 I=1,NSTEP,1    Step across x
        AI=I-1
        X=AI*ALEN/ANSTEP
        Y=1./3.
        DO 10 N=1,NHARM,1    Sum over harmonics
        AN=N
        A=(3./(PI**2*AN**2))*(1.-COS(2.*PI*AN/3.))    Coefficient
10      Y=Y+A*COS(2.*PI*AN*X/ALEN)    Amplitude
20      PRINT 100,X,Y
100     FORMAT(1X,2E15.5)
30      STOP
        END
```

Such a series converges quite rapidly, and after summing only a few harmonics, a good approximation of the function can be obtained [Programs 10.1(a), (b), (c)].

Figure 10.2 shows the results of summing this series with various numbers of harmonics. The function was plotted over a distance of $2L$ to demonstrate the periodicity of the series. The region over which you use the Fourier series to calculate the values of the function does not have to be the period you used to find the Fourier coefficient.

Various other periodic functions are also easy to write in a closed-form Fourier-series representation. Several such examples are the triangle, square, and sawtooth functions.

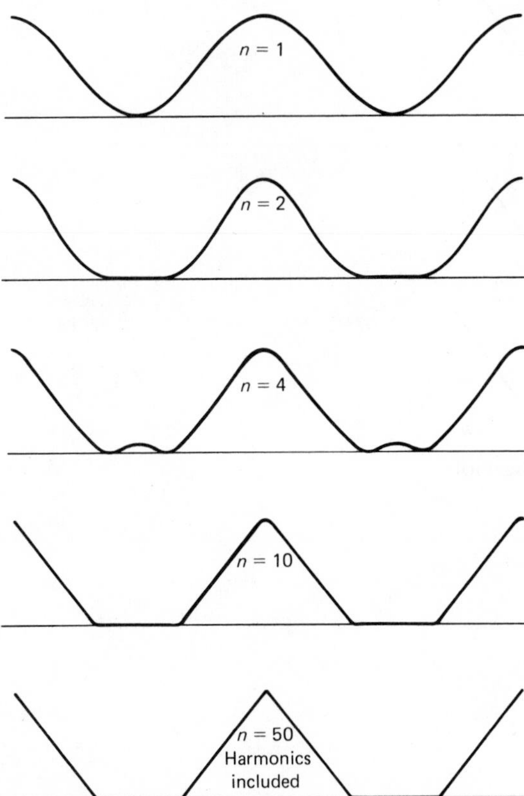

Figure 10.2 Summing various numbers of harmonics for the spiked wave.

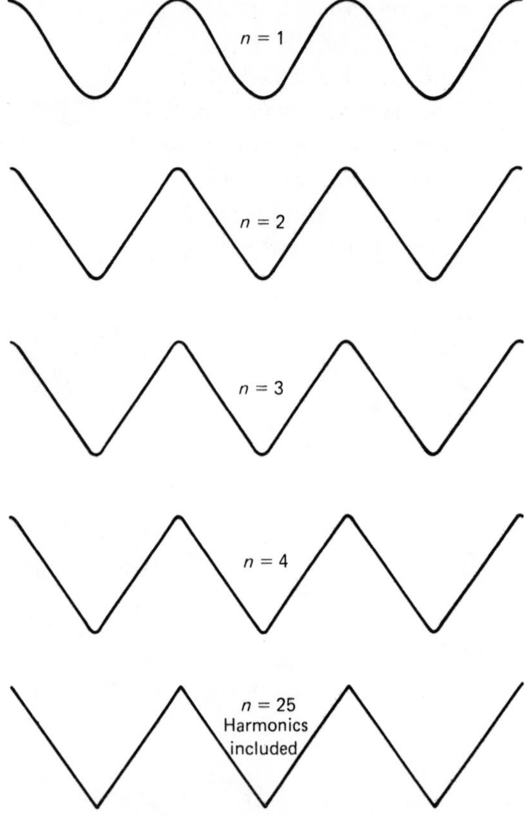

Figure 10.3 Summing various numbers of harmonics for the triangle wave.

Triangle function:

$$f(x) = \begin{cases} 1 + \dfrac{2}{L}x & \dfrac{-L}{2} < x < 0 \\ 1 - \dfrac{2}{L}x & 0 < x < \dfrac{L}{2} \end{cases}$$

Square function:

$$f(x) = \begin{cases} +1 & 0 < x < \dfrac{L}{2} \\ -1 & \dfrac{L}{2} < x < L \end{cases}$$

Sawtooth function:

$$f(x) = x \qquad 0 < x < L$$

It is instructive for the student who has never calculated Fourier coefficients to do so for the periodic functions above. The results from summing various numbers of harmonics for the waves are shown in Figures 10.3 to 10.5. In the cases of the square and sawtooth functions, you will notice an overshoot which occurs at the function's discontinuity. This overshoot is called the Gibbs phenomenon. In the limit of infinitely many terms, the overshoot remains finite. The program SQRWAVE [Programs 10.2(a), (b)] sums the harmonics for the square wave. On some computers, the calculation of sines and cosines is very slow. A fast way to calculate these functions is discussed in the Appendix to the book.

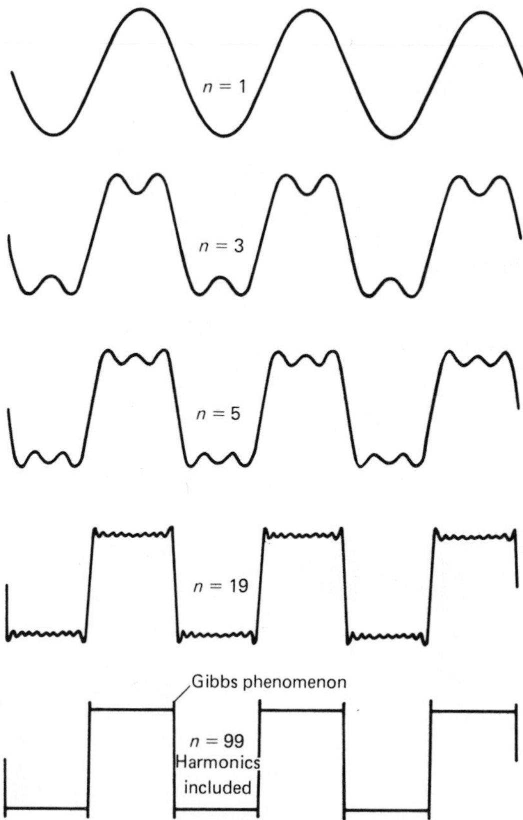

Figure 10.4 Summing various numbers of harmonics for the square wave.

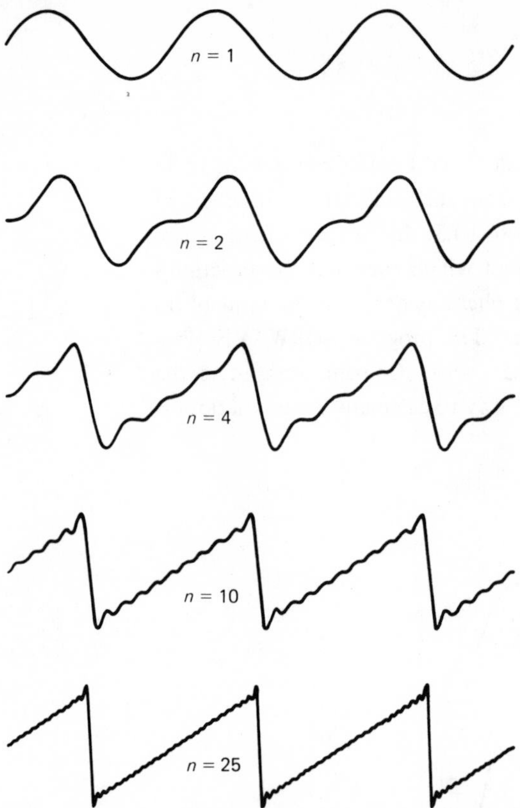

Figure 10.5 Summing various numbers of harmonics for the sawtooth wave.

Program 10.2 (a)

```
SQRBAS      Synthesis of square wave

100 LET P=3.14159     π
110 LET L=2*P     Period
120 LET H=50     Number of harmonics
130 DIM B(50)
140 FOR N=1 TO H                         } Store
150 LET B(N)=2/(P*N)*(1-COS(P*N))        } coefficients
160 NEXT N
170 PRINT "X","AMPLITUDE"
180 FOR X=0 TO L STEP L/10     Step across x
190 LET Y=0
200 FOR N=1 TO H     Sum over harmonics
210 LET Y=Y+B(N)*SIN(2*P*N*X/L)     Amplitude
220 NEXT N
230 PRINT X,Y
240 NEXT X
250 END
```

Program 10.2 (b)

```
SQRFOR     Synthesis of square wave

           DIMENSION B(50)
           PI=3.14159265
           WL=2.*PI    Period
           IH=50    Number of harmonics
           DO 5 N=1,IH                          } Store
5          B(N)=2./(PI*N)*(1-COS(PI*N))         } coefficients
           DO 15 IX=1,11    Step across x
           X=(IX-1)*WL/10.
           Y=0.
           DO 10 N=1,IH    Sum over harmonics
10         Y=Y+B(N)*SIN(2*PI*N*X/WL)    Amplitude
15         PRINT 100,X,Y
20         STOP
100        FORMAT(1X,2F10.4)
           END
```

It is not necessary to have a simple analytical expression for a function in order to determine the Fourier-series expansion. The program FSCOEFF [Programs 10.3(a), (b)] uses numerical integration to find the Fourier coefficients for any periodic function or any periodic set of data. As the program is now written, it will calculate the coefficients for the square wave pictured in Figure 10.4. The amplitude of the square wave, for equally spaced steps in x, is stored in a vector. The numerical integration uses the trapezoidal rule. The smaller the step in x, the greater the accuracy of the coefficients (especially the higher-order coefficients). By filling the vector with data points, any periodic function can be Fourier-analyzed and the coefficients can be determined. The numerical integration is easiest if the data points are equally spaced throughout the period.

There are a number of applications of Fourier analysis in data handling. You can perform a very beautiful set of experiments employing data handling. One experiment involves photographing an EKG or a musical tone displayed on an oscilloscope. The EKG can be your own; only the crudest instrumentation is necessary. The tone can be from any instrument you wish. The photograph is then reduced to a set of amplitudes at various times and the data are Fourier-analyzed. Since the pattern is periodic, the Fourier analysis is in terms of a Fourier series, and the mathematical manipulations are simply those discussed above. The resulting frequency spectrum has an obvious structure—fundamental components at the basic heartbeat frequency or at the fundamental note; higher components from movements of the various portions of the heart during each beat or from the harmonic content of the instrument (overtones). On top of all the real information is the noise pattern; but in Fourier frequency space, this noise is usually easy to pick out. You can, if you wish, resynthesize the Fourier pattern after eliminating the noise part. The result is an increased signal-to-noise ratio—the aim of any good experimentalist. The frequency spectrum information remains in the computer for any further data reduction.

Program 10.3 (a)

FSCOEBAS *Calculates Fourier series coefficients*

```
100 DIM F(101)
110 LET P=3.14159  π
120 LET N1=10  Number of coefficients to calculate
130 LET N=100  Number of data intervals
140 FOR I=1 TO N+1
150 READ F(I)
160 NEXT I
170 DATA .5,0,0,0,0,0,0,0,0,0,0,0,0,0,0,0,0,0,0,0,0,0,0,0,0
180 DATA 0,0,0,0,0,0,0,0,0,0,0,0,0,0,0,0,0,0,0,0,0,0,0,0,0
190 DATA .5,1,1,1,1,1,1,1,1,1,1,1,1,1,1,1,1,1,1,1,1,1,1,1,1
200 DATA 1,1,1,1,1,1,1,1,1,1,1,1,1,1,1,1,1,1,1,1,1,1,1,1,1,.5
                                        } Put in wave shape (one cycle)
210 LET A0=1/N*F(1)/2
220 FOR I=2 TO N
230 LET A0=A0+1/N*F(I)
240 NEXT I
250 LET A0=A0+1/N*F(N+1)/2
260 PRINT "A0=";2*A0
                                        } Zeroth coefficient (trapezoid rule)
270 PRINT "HARMONIC","A","B"
280 FOR J=1 TO N1  Step through coefficients
290 LET A=0
300 LET B=0
310 FOR I=1 TO N+1  Integrate over data
320 LET A=A+2/N*F(I)*COS(2*J*P*(I-1)/N)
330 LET B=B+2/N*F(I)*SIN(2*J*P*(I-1)/N)
340 NEXT I
350 LET A=A-1/N*F(1)-1/N*F(N+1)
                                        } Each coefficient in turn (trapezoid rule)
360 PRINT J,A,B
370 NEXT J
380 END
```

Program 10.3 (b)

FSCOEFOR *Calculates Fourier series coefficients*

```
      DIMENSION FNCT(101)
      PI=3.14159
      NUM=10  Number of coefficients to calculate
      INTRVL=100  Number of data intervals
      FINT=INTRVL
      DATA FNCT/.5,49*0.,.5,49*1.,.5/  Wave shape
      A0=1./FINT
      DO 10 LOC=2,INTRVL
10    A0=A0+1./FINT*FNCT(LOC+1)/2.
      A0=2.*A0
      PRINT 101, A0
                                        } Zeroth coefficient (trapezoid rule)
      DO 20 INDEX=1,NUM  Step through coefficients
      FINDEX=INDEX
      A=0.
      B=0.
      DO 30 LOC=1,INTRVL+1  Integrate over data
      FLOC=LOC
      A=A+2./FINT*FNCT(LOC)*COS(2.*FINDEX*PI*(FLOC-1.)/FINT)
30    B=B+2./FINT*FNCT(LOC)*SIN(2.*FINDEX*PI*(FLOC-1.)/FINT)
      A=A-1./FINT*FNCT(1)-1./FINT*FNCT(INTRVL+1)
20    PRINT 106,INDEX,A,B
                                        } Each coefficient in turn (trapezoid rule)
101   FORMAT(1X,3HA0=,F10.5)
106   FORMAT(I3,2F10.5)
      END
```

This laboratory experiment is a particularly nice project, especially for students who are medically or musically inclined. Clarinets, flutes, and guitars have been found to work nicely. You can even study tonguing and other transient phenomena.

Example An example of the analysis of data can be seen in Figure 10.6. The data (Table 10.1) are the voltages at various intervals for the heartbeat pictured. The resulting values of the Fourier coefficients for the first 45 harmonics are listed in the following table. These coefficients were determined by the program FSCOEFF.

The resynthesis of the wave is also pictured in Figure 10.6. With the first 25 harmonics, the structure of the wave is easily detectable; however, the electrical noise has been largely removed. With the first 100 harmonics, the original wave is almost exactly reproduced.

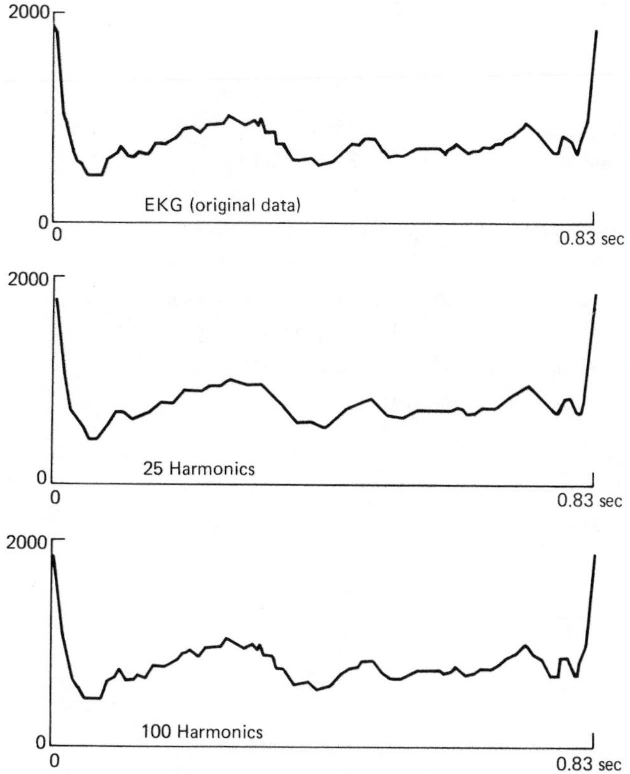

Figure 10.6 A plot of an EKG and its resynthesis with 25 and 100 harmonics. With 25 harmonics, most of the noise is removed. With 100 harmonics, the curve is almost exactly reproduced.

Table 10.1 Table of Fourier Coefficients for Heartbeat in Figure 10.6

n	A_n	B_n	n	A_n	B_n	n	A_n	B_n
1	47.5	39.4	16	33.8	19.5	31	5.2	10.3
2	0.8	-100.4	17	31.7	24.9	32	6.2	14.1
3	117.4	-53.4	18	24.3	15.6	33	0.2	7.3
4	31	28.2	19	23.3	20.6	34	2.5	4.3
5	88.5	-8.5	20	21.8	8.8	35	2.4	4.4
6	60.6	46.3	21	19.9	10.9	36	5.8	0.9
7	102.2	19.6	22	8.7	9.2	37	5	5.5
8	72.2	16.9	23	12.7	2.6	38	3.6	6.3
9	63.2	36.3	24	15.3	6.1	39	6.2	7.7
10	48.7	21.9	25	10	2.5	40	0.1	7.2
11	21.2	30.9	26	-0.5	3.1	41	2.9	1.6
12	27.8	25.8	27	4.6	6.3	42	-0.9	5.2
13	20.7	26.8	28	-0.8	11.9	43	0.3	-0.6
14	16.6	20.3	29	0.6	3.6	44	0.8	3.3
15	37.4	22.8	30	3.5	11.7	45	3.5	-1.9

10.4 Fourier Transforms

Most of the waves in nature are pulses of one sort or another. By Fourier's theorem, a pulse can be considered to be the sum of an infinite number of different frequency components. The components are sines and cosines closely spaced in frequency. For example, consider a triangular pulse symmetric around $x = 0$ (Figure 10.7). When you transform this pulse into Fourier space (also called k space or wave vector space), you will find only cosine waves. This is because $f(x) = f(-x)$; the pulse has even symmetry. Cosines have even symmetry, whereas sines have odd symmetry. You would be surprised to find odd-symmetry components in an even-symmetry pulse. In fact, this triangular pulse has no sine components.

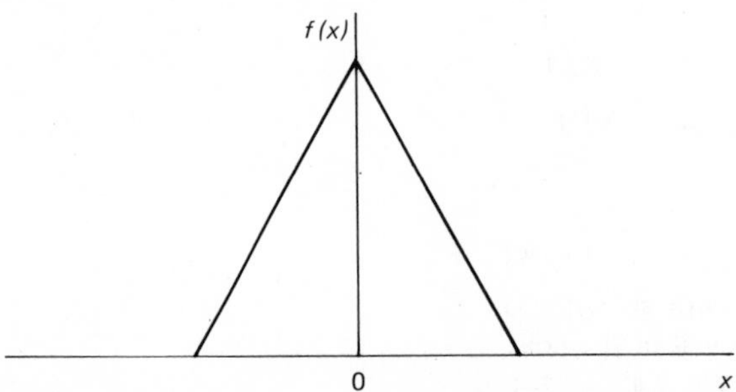

Figure 10.7 Triangular pulse symmetric around zero.

You find the coefficients of all the cosine waves by the integral

$$a(k) = \frac{1}{2\pi} \int_{-\infty}^{\infty} f(x) \cos(kx)\, dx$$

Every value of k will have an amplitude $a(k)$; $a(0)$ is the largest amplitude for the triangular pulse; $a(k)$ is essentially the amount of $\cos(kx)$ in the original pulse.

10.5 Strategy for Dispersive Waves

When using Fourier transforms, we find that wave propagation is straightforward. It is practically always easiest to find the propagation properties of sine and cosine waves in any medium. Each cosine (or sine) wave has its own velocity $v(k)$ [and absorption constant $A(k)$, in general].

The program WAVES [Programs 10.4(a), (b), (c)] gives one way to perform wave-propagation calculations. WAVES first calculates the Fourier transform of the initial pulse shape. The program then propagates each wave component k according to the correct velocity $v(k)$ for some amount of time. WAVES then resynthesizes (puts back together) the pulse using the same amounts $a(k)$ of the components that were present initially. Using the same coefficients $a(k)$ implies that the wave components do not change size—they are not absorbed and one component does not turn into another.

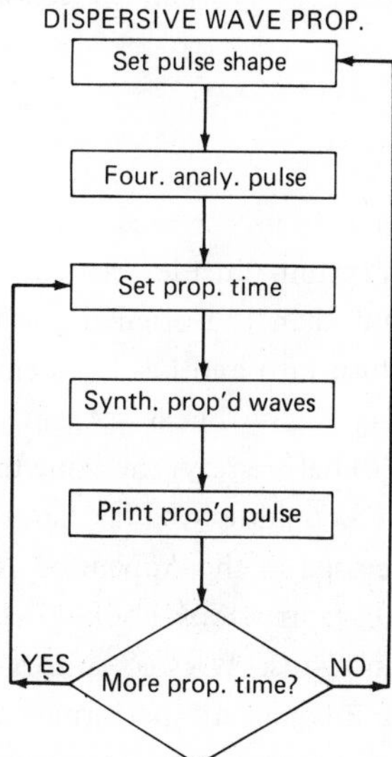

Program 10.4 (a) Flow chart for a dispersive-wave-propagation strategy. After setting the shape of the pulse propagated, the strategy adds up the sum of the Fourier components at each point.

Program 10.4 (b)

```
WAVESBAS     Propagates wave pulse dispersively

100 DIM F(161),G(161)
110 FOR I=1 TO 161     Store Fourier amplitudes of components
120 LET K0=-5*3.14159+(I-1)*3.14159/16
130 LET F(I)=1.41421*EXP(-K0*K0)     Real part
140 IF F(I)>1E-12 THEN 160     } Protect against
150 LET F(I)=0                 } underflow
160 LET G(I)=0
170 NEXT I
180 DIM C(360),S(360)
190 LET P0=6.28318/360
200 FOR I=1 TO 360             } Store sines
210 LET C(I)=COS((I-1)*P0)     } and cosines
220 LET S(I)=SIN((I-1)*P0)
230 NEXT I
240 PRINT "TIME";
250 INPUT T1     Time of propagation
260 PRINT "X","AMPLITUDE↑2"
270 FOR X1=-5 TO 5 STEP .5     Step across x
280 LET I1=0
290 LET I2=0
300 FOR I=1 TO 161     Sum over components
310 LET K0=-5*3.14159+(I-1)*3.14159/16
320 LET V0=1+.01*K0     Velocity at wave vector, k0
330 LET K9=K0*(X1-V0*T1)                              } Find correct entry in
340 LET K8=1+INT((K9-INT(K9/6.28318)*6.28318)/P0)     } C( ) and S( )
350 LET R1=C(K8)*F(I)+S(K8)*G(I)     Real part of synthesis
360 LET R2=C(K8)*G(I)-S(K8)*F(I)     Imaginary part
370 LET I1=I1+R1
380 LET I2=I2+R2
390 IF (I-1)*(I-161)<>0 THEN 420     } Correct end
400 LET I1=I1-R1/2                   } points of trapezoid
410 LET I2=I2-R2/2                   } rule
420 NEXT I
430 LET I1=I1*3.14159/16/SQR(6.28318)     } Normalize real
440 LET I2=I2*3.14159/16/SQR(6.28318)     } and imaginary parts
450 PRINT X1,I1*I1+I2*I2
460 NEXT X1
470 PRINT
480 GOTO 240     Return for new propagation time
490 END
```

[The absorption demand is not serious; you can write your own program in such a way as to include absorption. Just let the coefficients in the resynthesis be smaller by the appropriate amount than the original coefficients. The approximation that one component does not turn into another is a demand that the medium be "linear." Almost all media are very highly linear. Again you can allow for nonlinearity, if you wish, by mixing the $a(k)$ before resynthesizing the pulse.]

The program WAVES could use the fast sines and cosines mentioned in the Fourier-series section and discussed in the Appendix. Another more complicated way that is discussed in numerical spectroscopy books to speed Fourier-transform calculations is called fast Fourier transforms. WAVES approximates the infinite integrals involved in Fourier analysis. The integrals are performed from -5 to $+5$ in steps of $\frac{1}{16}\pi$. These values were chosen because they seem to calculate the infinite integral reasonably cor-

Program 10.4 (c)

```
WAVESFOR     Propagates wave pulse dispersively

          DIMENSION REPULS(161),REFOUR(161),AIFOUR(161)
          PI=3.14159
          DO 20 IK=1,161,1    Store Fourier amplitudes of components
          AIK=IK-1
          AK=-5.*PI+AIK*PI/16.
          REFOUR(IK)=1.41421*EXP(-AK*AK)    Real part of Fourier amplitude
          IF (REFOUR(IK)-1.E-12) 15,20,20
15        REFOUR(IK)=0.
20        AIFOUR(IK)=0.
30        READ(5,100)TPROPD    Time pulse has propagated
          IF (TPROPD.EQ.999.) GOTO 60
          DO 50 IX=1,21    Step across x
          AIX=IX-1
          X=-5.+AIX*.5
          RESYN=0.
          AIMSYN=0.
          DO 40 IK=1,161    Sum over components
          AIK=IK-1
          AK=-5.*PI+AIK*PI/16.
          VELATK=1.+.01*AK    Velocity at wave vector
          CSSYN=COS(AK*(X-VELATK*TPROPD))
          SNSYN=SIN(AK*(X-VELATK*TPROPD))
          RESYN=RESYN+ Q.25331/16.)*(CSSYN*REFOUR(IK)+SNSYN*AIFOUR(IK))    Real part
40        AIMSYN=AIMSYN+(1.25331/16.)*(CSSYN*AIFOUR(IK)-SNSYN*REFOUR(IK))    Imaginary part
          AMPL=RESYN**2+AIMSYN**2
50        PRINT 101,X,AMPL
          GOTO 30    Return for new propagation time
100       FORMAT(E5.2)
101       FORMAT(1X,2E15.5)
60        STOP
          END
```

rectly while minimizing the amount of computer time. The integrals are performed by the trapezoidal rule.

Figure 10.8 shows the propagation of a triangular pulse in a nondispersive medium. The first parts of the figure show the original pulse and its Fourier spectrum; the latter parts show propagation of the pulse forward and backward in time. The pulse moves as a whole, since the velocities of all the components of the wave are the same. The figure is a reminder that when you do a complicated calculation, you can check the calculation in a simple situation—one in which you already know the answer.

Figure 10.9 shows the same triangular pulse propagated in a slightly dispersive medium. The velocity for waves traveling in this medium has the form

$$v(k) = v_0\,(1 + 0.01\,k)$$

The velocity depends slightly on which component k of the wave is being propagated. When the components are resynthesized, some have moved further than others and the result is a distortion of the pulse shape. The distortion increases as the pulse moves along. The group velocity of the pulse is defined to be the rate at which the center of

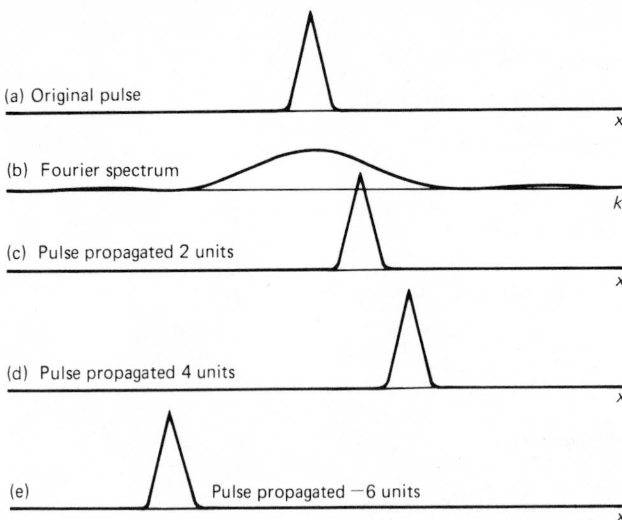

Figure 10.8 Propagation of a triangular pulse through a nondispersive medium. The original pulse and its Fourier transform are shown. The pulses after 2, 4, and −6 units of time are also shown. The similarity in the pulse shapes gives the student confidence in the simulation method.

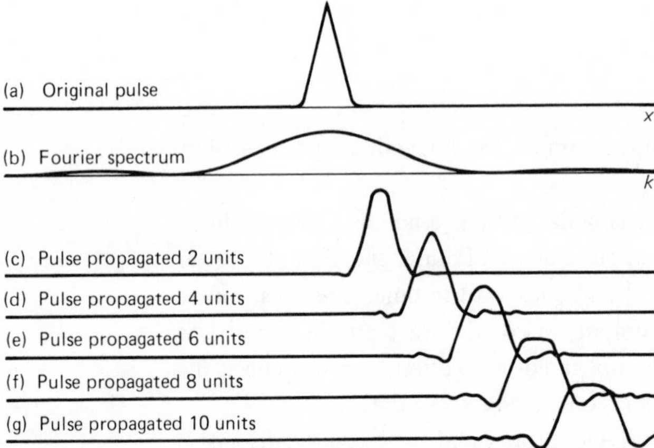

Figure 10.9 Propagation of a triangular pulse through a slightly dispersive medium. (a) The original pulse; (b) the real part of the Fourier transform of the initial pulse; (c) through (g) the pulse after 2, 4, 6, 8, and 10 units of time have passed. The distortion of the pulse is due to the dispersive nature of the propagation medium. The zero of spatial Fourier-transform coordinates is at the center of the figure; the left and right sides of the figure correspond to -4π and $+4\pi$, respectively. All the figures are scaled this way. Curves are separated vertically for clarity.

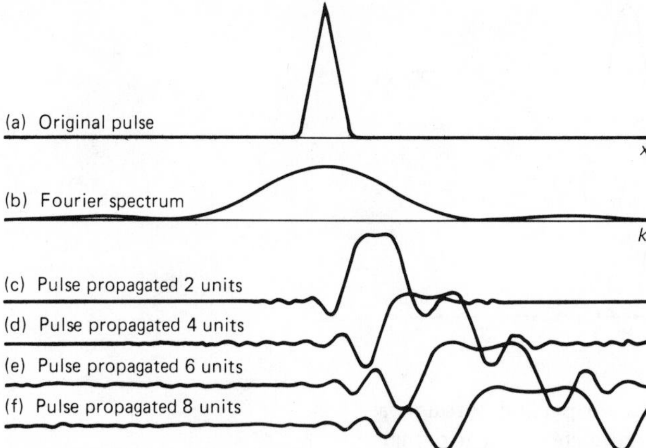

Figure 10.10 Propagation of a triangular pulse through a strongly dispersive medium. (a) The original pulse; (b) the real part of the Fourier transform; (c) through (f) the pulse after propagating 2, 4, 6, and 8 units of time.

the pulse (or the pulse envelope) progresses. This group velocity may not be the same as the velocity of any of the wave components. You could measure the group velocity from the figure.

Figure 10.10 shows the same pulse moving through a medium with heavier dispersion, $v(k) = v_0\ (1 + 0.1k)$. These two dispersive velocities are models of the way in which a light pulse propagates near the edges of optical absorption bands in solids and gases.

So far you have dealt only with triangular pulses. Figure 10.11 shows a pulse of $2\frac{1}{2}$ cycles of a cosine wave and its Fourier transform. Notice that because the cosine wave is cut off into a pulse, many cosine components are needed to form the pulse. Nonetheless, the Fourier spectrum peaks around the basic cosine wave; as the pulse contains more and more cycles, the Fourier spectrum peaks more and more sharply around the basic cosine. This makes physical sense because the Fourier spectrum for an infinitely long cosine train is just a spike at the basic cosine component.

Figure 10.11(b) and (c) demonstrate another fact about dispersive media. When the pulse is resynthesized after some propagation time, some sine components may appear. Figure 10.11(a) is the initial pulse and its Fourier transform. Figure 10.11(b) shows the cosine part of the pulse after several propagation times and Figure 10.11(c) shows the sine part for the same times. The distortion of the pulse is obvious as is the phase mixing in a dispersive medium. The velocity for this figure is that of Figure 10.10, $v = v_0\ (1 + 10.1k)$.

Figure 10.12(a), (b), (c) suggests an interesting application of these dispersive-wave methods to solid-state physics. It turns out that elastic waves (phonons) propagating

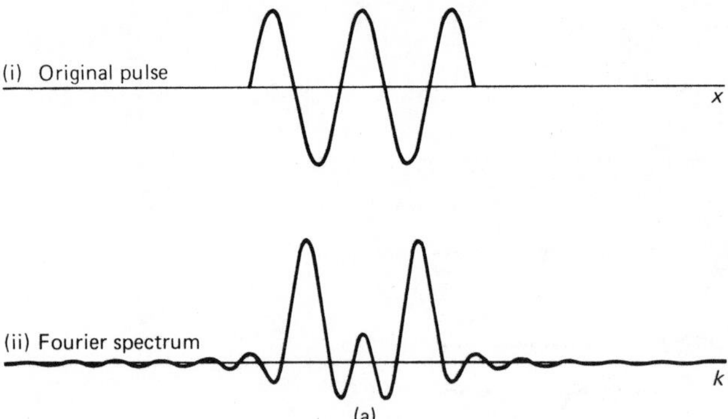

Figure 10.11 (a) Propagation of a wave packed through a strongly dispersive medium. This figure shows the original pulse and the real part of its Fourier transform. The Fourier spectrum is peaked near plus and minus the wave vector of the cosine part of the pulse shape.

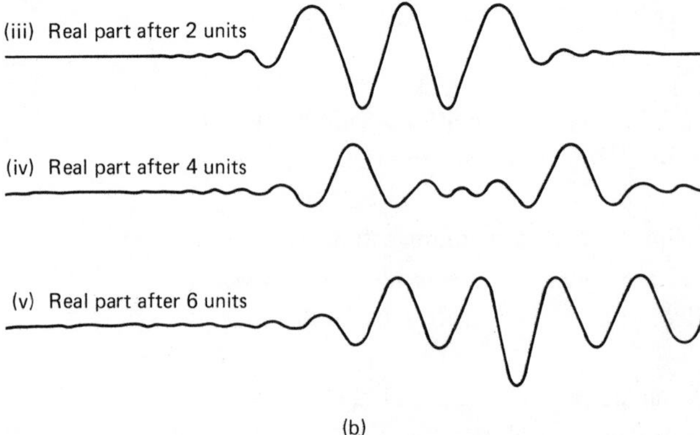

Figure 10.11 (b) Propagation of a wave packet through a strongly dispersive medium. This figure shows the real or cosine part of the pulse after propagating 2, 4, and 6 units of time.

through a (one-dimensional) solid have the phase velocity $v(k) = v_0 \sin(ka/2)$, where a is the distance between the atoms in the solid and $k = 2\pi$ divided by the wavelength. When one-half the wavelength of the wave is the same as this lattice spacing a, the wave does not get anywhere even though it tries $[ka = \pi$ so $v(k) = 0]$. Figure 10.12(a) shows a pulse with a wavelength much longer than a; the wave pulse moves but distorts a little. Figure 10.12 (b) has one-half the wavelength of the pulse closer to the lattice spacing;

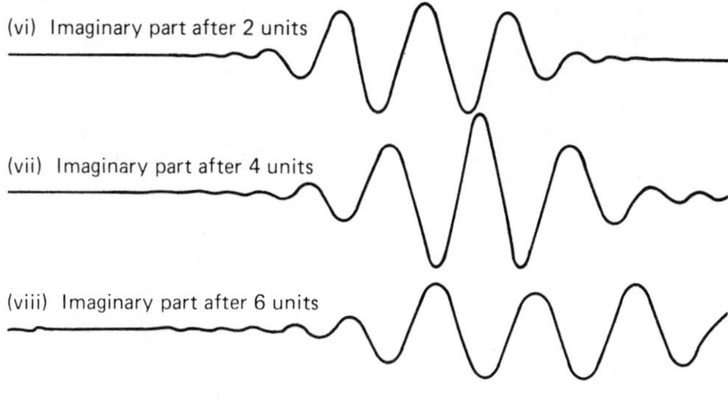

(c)

Figure 10.11 (c) Propagation of a wave packet through a strongly dispersive medium. This figure shows the imaginary or sine part of the pulse after propagating 2, 4, and 6 units of time.

the wave distorts more rapidly, and the group velocity is noticeably smaller. Figure 10.12(c) shows the propagation of a pulse whose basic cosine wavelength is just barely larger than twice the lattice spacing a. The distortion of the pulse is obvious, as is the fact that the group velocity of the pulse (the velocity of the envelope of the pulse) is nearly zero! These are the sorts of effects you find in real wave-propagation situations.

Problems

The following are representative of exercises students have done using the material developed in this chapter.

1. Using the Fourier-series results quoted in the chapter, write a program summing 1, 5, 10, and 100 terms for each of the following:
   a. The triangle function
   b. The sawtooth function
   c. The square function

   Point out and explain the Gibbs phenomenon.

2. a. Look up the Fourier transform of a square pulse.
   b. Write a program to propagate a square pulse through a medium such that $v(k) = k^2$. [Check your program first by using $v(k)$ = a constant.]
   c. Propagate the pulse both forward and backward in time. Explain your results physically.

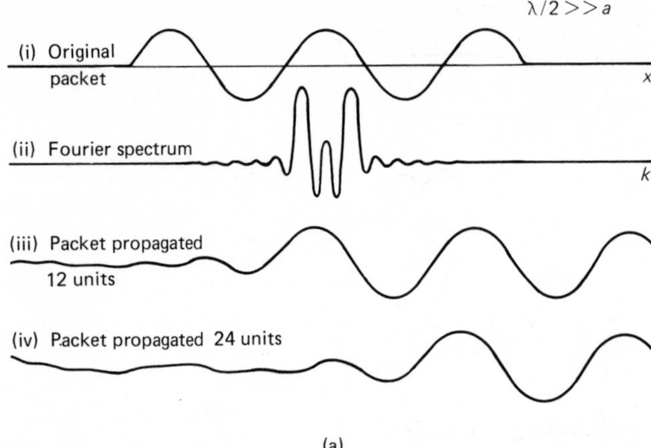

Figure 10.12 (a) Propagation of a phonon wave packet through a one-dimensional lattice. This figure shows the initial pulse, its Fourier spectrum, and its propagation for 12 and 24 time units. This pulse has a wavelength substantially longer than the lattice spacing, so distortion is minimal.

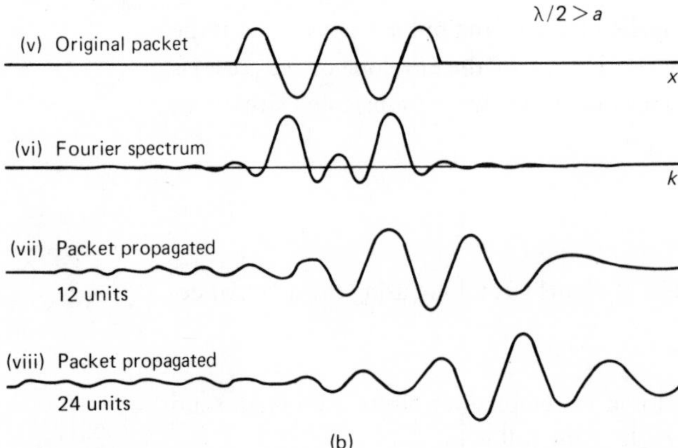

Figure 10.12 (b) Propagation of a phonon wave packet through a one-dimensional lattice. This figure shows the effects on a pulse whose basic wavelength is nearer twice the lattice spacing. The distortion is more noticeable and the group velocity is smaller.

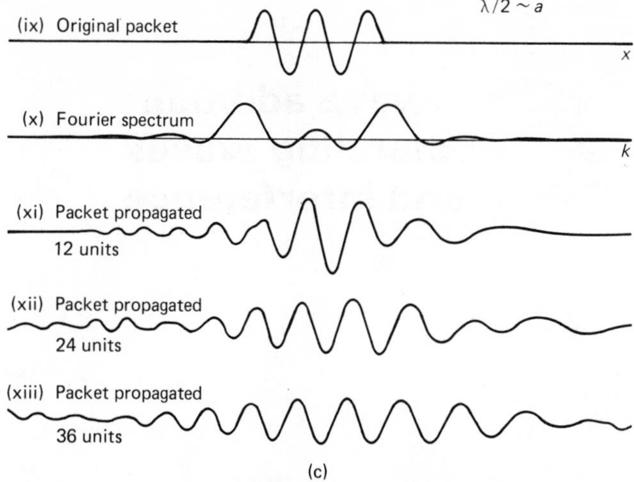

Figure 10.12 (c) Propagation of a phonon wave packet through a one-dimensional lattice. This figure shows the effects on a pulse whose basic wavelength is nearly twice the lattice spacing. The distortion is large, and the group velocity is nearly zero.

3. a. Take photographs (using an oscilloscope, a microphone, and a scope camera) of the amplitude versus time for several notes played by an instrument of your choice. Be sure each photograph has $1\frac{1}{2}$ to 2 cycles of the repetitive waveform.
   b. From your photograph, read off the amplitudes at 100 equally spaced times across one cycle (101 points including both ends of the cycle). Enter these amplitudes as data in a program which calculates Fourier coefficients. What is the frequency of each note?
   c. Find the first 10 Fourier coefficients for the waveform from each photograph.
   d. Look carefully at the harmonic content of each note. Try to relate the harmonic content to the timbre of the note. If you know someone doing this problem with a different instrument, compare results.

4. It is possible to have the computer perform the Fourier analysis of any pulse shape into its Fourier spectrum [by performing numerical integrations of $\int_{-\infty}^{\infty} f(x) \cos(kx)\, dx$, and so forth]. Write such a Fourier analysis program and Fourier-analyze some interesting pulse shape that can not be handled analytically. (Check your program first on the triangle pulse whose analytic Fourier spectrum is known.)

11 wave addition: standing waves and interference

11.1 Introduction

The addition of waves leads to some very interesting and very important physical effects. Since waves occur throughout the physical world, effects such as resonance, interference, and diffraction are often part of your everyday experience. The pattern of sound around a hi-fi speaker is usually dominated by diffraction effects; the cross of light from a streetlight shining through a window screen is due to diffraction; and the sound from an organ pipe is a result of standing waves. The pattern of waves on the top of a vibrating glass of water is an interference effect; sound waves at the resonance frequency for a goblet can shatter glass.

Waves are added by summing the amplitudes. If two waves of the same frequency and wavelength reach the same point in space at the same time [one wave $A_1 \cos(kx + \omega t + \theta_1)$; the second wave $A_2 \cos(kx + \omega t + \theta_2)$], the resultant wave is simply $A = A_1 \cos(kx + \omega t + \theta_1) + A_2 \cos(kx + \omega t + \theta_2)$. Interesting wave-addition effects occur when the phases θ_1, θ_2, are such that the waves cancel or aid each other. When you sense this resultant wave, you practically always measure the intensity (the time average of the square of the amplitude) of the resultant wave. Interference occurs because of the cross term in the square of the amplitude A^2. Any number of waves may interfere at any point, and the resulting intensity pattern after interference can become quite complicated. Standing waves occur when a series of waves from the same source continuously reach the same point with the correct phase differences.

11.2 Standing Waves

Standing waves are everywhere in nature. Violin strings, organ pipes, and guitars are all examples of standing waves. So is the collapse of the Tacoma Narrows Bridge. The general theory of standing waves is quite complicated, especially if you want to keep track of wave losses. The basic ideas behind standing waves are easy to understand.

The results of one-dimensional standing-wave calculations predict resonant responses of the wave system at definite frequencies ω and wavelengths λ. For example, if both ends of a string are fixed (or an organ pipe has both ends closed), you expect resonances at $\lambda = 2L, L, \frac{2}{3}L, \frac{1}{2}L, \ldots$, where L is the length of the string or pipe. The reflection of waves at a closed end involves a 180° phase change for the wave; the standing-wave pattern has a node at such an end.

If one end of the resonant chamber is closed and one is open, you expect resonances at $\lambda = 4L, \frac{4}{3}L, \frac{4}{5}L, \frac{4}{7}L, \ldots$, because the reflection of a wave at an open end has no phase change associated with it. If both ends of the chamber are open, you still expect resonances. The resonances are very similar to those with both ends closed, except that loops (antinodes) instead of nodes of the wave occur at the (open) ends.

These simple pictures of standing waves assume no loss of wave amplitude in the medium and perfect reflections at the ends. Neither of these assumptions is true for a real-wave situation, but both are reasonable approximations in many cases.

The computer method, on the other hand, can be applied to waves reflecting back and forth between the ends of the chamber with realistic boundary conditions. A source emits waves which reach some observation point directly. These waves also continue to the end of the chamber and (partially) reflect. This first reflected wave reaches the observation point and adds to the original wave. The wave continues to reflect off the two ends of the chamber, and the total wave response at the observation point is the sum of all the waves, direct as well as reflected. The final amount of the wave observed is the sum of a series. If the wave speed is v and the wave frequency is f, then the wave vector $k = 2\pi/\lambda = 2\pi f/v$. With the original source at a point x_9 and an observation at x_0, the direct wave is $A \cos [k(x_0 - x_9)]$, where A is the original amplitude. {A loss in amplitude due to absorption by the medium can be included by changing the amplitude. For example, the wave might become the original wave with amplitude $A \exp [-\alpha(x_0 - x_9)]$.} When the first reflected wave returns to the observation point, it has a different amplitude [$A_{new} = (A_{old})$(reflection coefficient)] and a different phase $\{P_{new} = P_{old} + [2k(x_{end} - x_0)]\}$. We have used the fact that the wave has made two trips—one down to the end and one back. The second reflected wave makes similar changes but is reflected off the other end.

The program STAND [Programs 11.1(a), (b), (c)] implements this computer strategy using sine and cosine functions. Sine and cosine functions use a large amount of computer time. Fast ways to calculate these functions are discussed in the Appendix.

The result of the program is the amplitude at the observation point after multiple reflections off the ends. You can run the program for a range of frequencies and demonstrate resonance curves.

Program 11.1 (a) Flow chart for a standing-wave strategy. After setting the values of initial variables, the strategy calculates the sum of the waves bounced off the ends of the region.

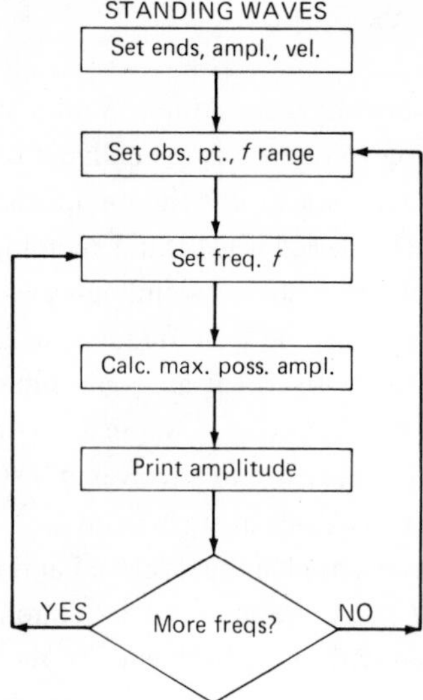

Program 11.1 (b)

```
STANDBAS      Standing waves

100 READ X1,R1,X2,R2,X9,A,V    First end, first reflection coefficient, second end, second
110 DATA 0,-.8,.165,-.8,0,1,330     reflection coefficient, position of source,
120 LET I0=-7.6/LOG(ABS(R1*R2))     amplitude, speed of waves
130 PRINT "ENDS OF PIPE:";X1;X2
140 PRINT "SPEED OF WAVE =";V
150 PRINT "OBSERVATION POINT";
160 INPUT X0
170 PRINT "LOWER FREQ., UPPER FREQ.";
180 INPUT F1,F2
190 PRINT "F","LAMBDA","AMPLITUDE"
200 FOR F=F1 TO F2 STEP (F2-F1)/10    Step across frequency
210 LET W=2*3.14159*F   Angular frequency
220 LET K=W/V   Wave vector
230 LET A0=0
240 LET T0=3.14159/W   Period
250 FOR T=0 TO T0 STEP T0/18   Step through full period
260 LET A8=A
270 LET P8=K*(X0-X9)-W*T   Phase at obs. pt.
280 LET A1=A*COS(P8)   Amplitude at obs. pt.
290 FOR I=1 TO I0   Sum over reflections
300 LET A8=A8*R2
310 LET P8=P8+2*K*(X2-X0)   Phase at second end
320 LET A1=A1+A8*COS(P8)   Amplitude after second end
330 LET A8=A8*R1
340 LET P8=P8+2*K*(X0-X1)   Phase at first end
350 LET A1=A1+A8*COS(P8)   Amplitude after first end
360 NEXT I
370 IF ABS(A1)<A0 THEN 390  }
380 LET A0=ABS(A1)          } Find maximum amplitude
390 NEXT T                  }
400 PRINT F,2*3.14159/K,A0
410 NEXT F
420 GOTO 170   Return for new frequency limits
430 END
```

Program 11.1 (c)

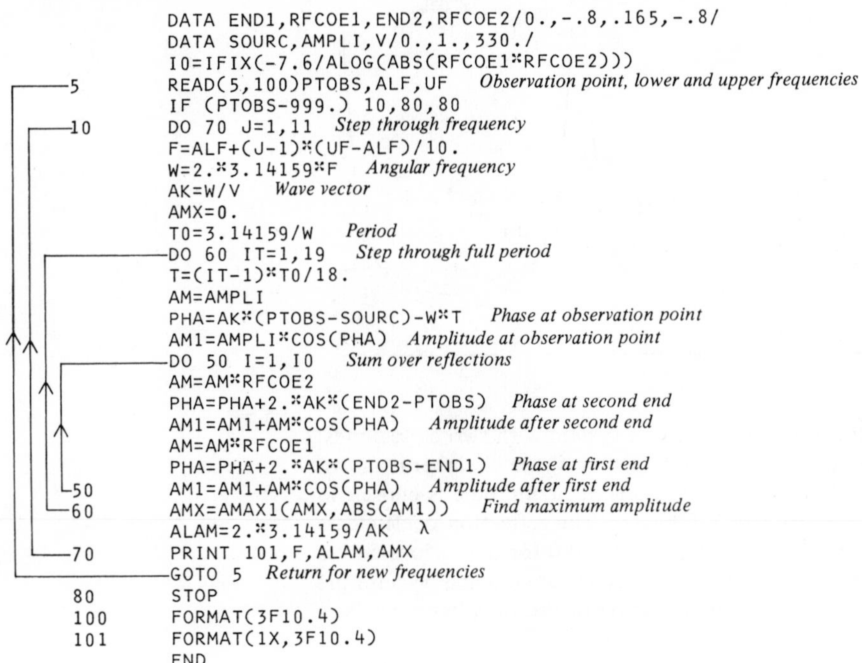

```
STANDFOR    Standing waves

         DATA END1,RFCOE1,END2,RFCOE2/0.,-.8,.165,-.8/
         DATA SOURC,AMPLI,V/0.,1.,330./
         IO=IFIX(-7.6/ALOG(ABS(RFCOE1*RFCOE2)))
5        READ(5,100)PTOBS,ALF,UF    Observation point, lower and upper frequencies
         IF (PTOBS-999.) 10,80,80
10       DO 70 J=1,11   Step through frequency
         F=ALF+(J-1)*(UF-ALF)/10.
         W=2.*3.14159*F   Angular frequency
         AK=W/V    Wave vector
         AMX=0.
         TO=3.14159/W    Period
         DO 60 IT=1,19   Step through full period
         T=(IT-1)*TO/18.
         AM=AMPLI
         PHA=AK*(PTOBS-SOURC)-W*T   Phase at observation point
         AM1=AMPLI*COS(PHA)   Amplitude at observation point
         DO 50 I=1,IO    Sum over reflections
         AM=AM*RFCOE2
         PHA=PHA+2.*AK*(END2-PTOBS)    Phase at second end
         AM1=AM1+AM*COS(PHA)    Amplitude after second end
         AM=AM*RFCOE1
         PHA=PHA+2.*AK*(PTOBS-END1)    Phase at first end
50       AM1=AM1+AM*COS(PHA)    Amplitude after first end
60       AMX=AMAX1(AMX,ABS(AM1))    Find maximum amplitude
         ALAM=2.*3.14159/AK    λ
70       PRINT 101,F,ALAM,AMX
         GOTO 5   Return for new frequencies
80       STOP
100      FORMAT(3F10.4)
101      FORMAT(1X,3F10.4)
         END
```

Figure 11.1 shows resonance effects in an organ pipe with two closed ends. For the sharpest resonances shown, each reflection loses 20% of the wave. Each reflection also reverses the phase by 180°. The other curves in Figure 11.1 show reflection coefficients of -0.6, -0.4, -0.2, and 0. The length is correct for a 1,000-Hz organ pipe, so you can expect resonances at 1000, 2000, . . . Hz. However, if the observation is at the center of the pipe, as in Figure 11.1, you will observe large resonant amplitudes only for odd modes; the even modes (for which 2, 4, 6, . . . , half-waves fit the chamber) have nearly zero amplitude at the center. Figure 11.2 shows the same two closed-end system, but now the observation point is one-eighth of the way down the tube. You observe all the resonances which occur up to 5000 Hz.

Figure 11.3 shows the resonances in a pipe with one open and one closed end; the observation point is at the open end. You see all the resonances. As the reflections at the ends become more nearly perfect, the resonances get sharper and larger. You still observe some resonant response off the resonant frequency. One number often used to specify the quality or sharpness of a resonance is Q = (the resonant frequency)/(the width of the resonance halfway up in amplitude). More nearly perfect reflections at the ends produce higher Q resonances.

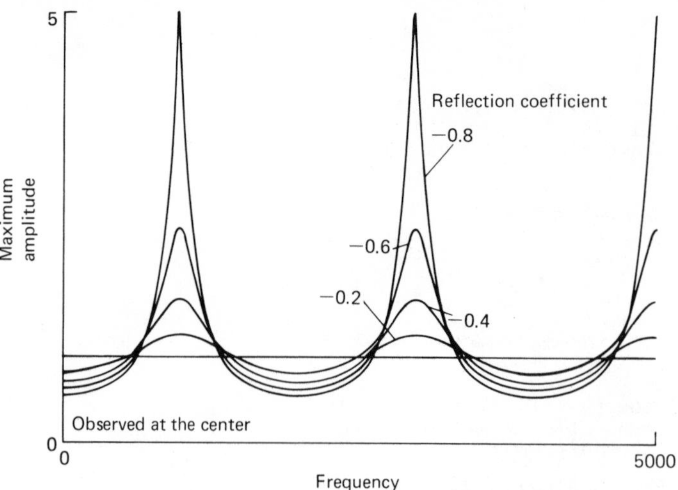

Figure 11.1 Resonant response of a system with two closed ends. The amplitude of the wave at the center of the chamber is shown as a function of frequency. The length of the pipe is chosen so that the pipe resonates at 1000 Hz. The reflection coefficients at the ends are −0.8, −0.6, −0.4, −0.2, and 0 for the various curves shown. The modes at 2000 and 4000 Hz are missing because the amplitude of those modes at the center of the chamber is small.

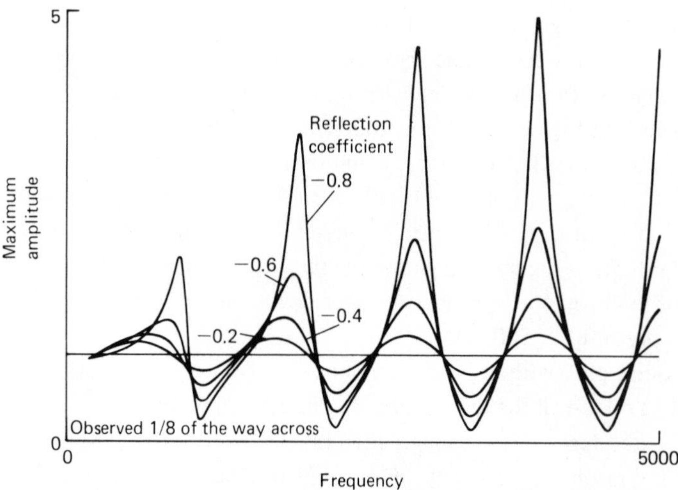

Figure 11.2 Resonant response of the chamber used for Figure 11.1. The observation point is one-eighth of the way down the tube, so all the resonances below 5000 Hz are observed.

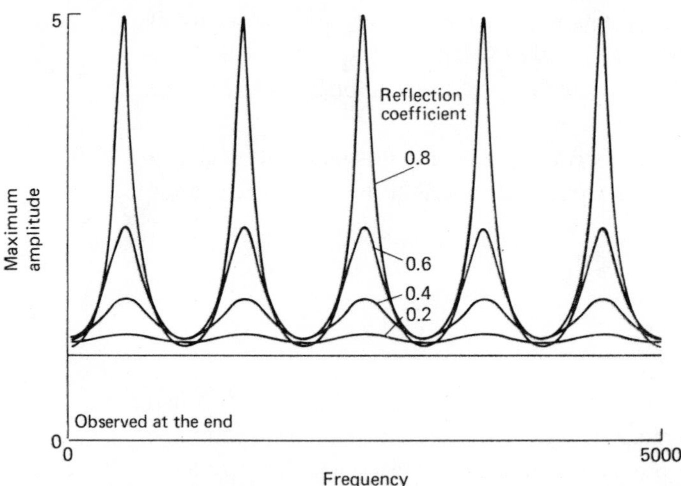

Figure 11.3 The resonant response of a chamber with one open and one closed end. The observation point is at the open end; results with reflection coefficients of ±0.8, ±0.6, ±0.4, ±0.2, and 0 are shown.

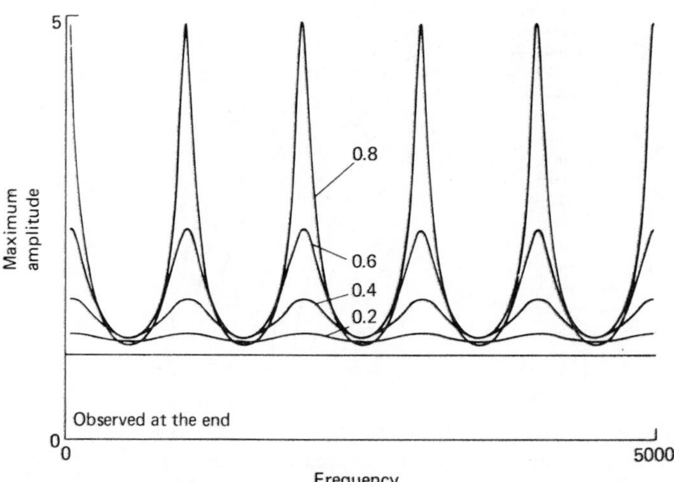

Figure 11.4 The resonant response of a chamber with both ends open. The resonances are observed at the open end; the resonance at 0 frequency is the trivial solution. Response curves with reflection coefficients of ±0.8, ±0.6, ±0.4, ±0.2, and 0 are shown.

Figure 11.4 shows resonances from a pipe with two open ends. The resonance at zero frequency is unphysical; it is called the trivial solution. You will also find this trivial solution if you solve the differential equation for the situation analytically. The other resonances are as expected.

This standing-wave computer method is entirely general. We could have included loss of the wave in the medium or complex phase reflections at the ends. The method is also easy to understand physically.

11.3 Huygens' Construction

Huygens was one of the first physicists to suggest that waves could be handled in a simple geometrical way. Huygens' principle says that all points on a wavefront can be considered as point sources for the production of secondary wavelets. After a time t, the new position of the wavefront will be the surface of tangency to these secondary wavelets. Although an approximation, the principle is of considerable use in optics.

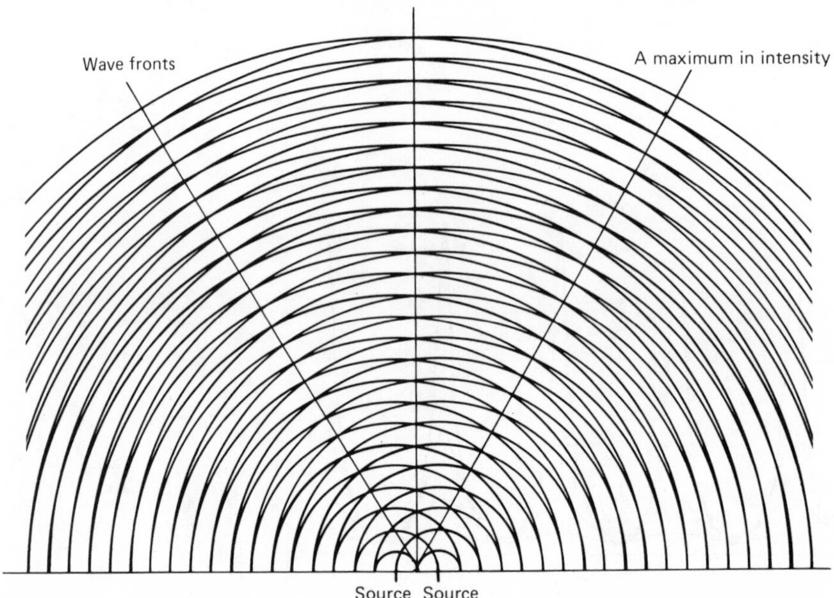

Figure 11.5 Huygens' construction for two-source interference. Only the forward directions are shown. The wavelength is half the source separation, so maxima occur at 0, $\pm 30^\circ$ and $\pm 90^\circ$ from the forward direction. The 0 and $\pm 30^\circ$ directions are drawn in. The directions in which the wavefronts are tangent are the directions of maximum intensity in the pattern. Notice that the angular widths of the maxima around $\pm 90^\circ$ are larger than those around the other directions.

In geometrical optics, the derivation of the law of refraction, using Huygens' principle, illustrates the fact that the velocity of light in a medium is less than that in vacuum and motivates the definition of the index of refraction as $n = c/v$. In physical optics, Huygens' construction gives a calculational method to deal with obstacles in the paths of waves. The method of secondary wavelets can be easily programmed. Such a program would calculate the contours of constant amplitude of the waves at some instant of time. The result would be a picture of the wave pattern at that instant (Figure 11.5). The lines of maximum intensity appear as the directions in which waves from the various secondary sources are moving together in phase. Having seen one or two such patterns, you can plot your own with a compass and a ruler.

11.4 Intensity Distribution

More often, though, interference (or diffraction) is visualized in terms of the intensity pattern across some screen placed in the wave field, or, more generally, in terms of a polar plot of intensity versus angle around the sources. You will see in a moment how to relate these various types of diagrams and how to use the computer to produce general intensity patterns for general wave-source distributions.

11.5 Basic Facts

The following basic facts are derived in your course text. If something looks new, review your textbook.

1. Reflection coefficient: Amplitude reflected/amplitude incident
   Transmission coefficient: Amplitude transmitted/amplitude incident
2. Reflection at a closed end includes a 180° phase change
   Reflection at an open end includes no phase change
3. Double-slit interference *(Young's experiment)*

   $I(\theta) = 4I_0 \cos^2 \beta$

   $\beta = \pi d \sin \theta / \lambda$

   Maxima at $d \sin \theta = m\lambda, \quad m = 0, 1, 2 \ldots$ *(Constructive interference)*
   Minima at $d \sin \theta = (m + \frac{1}{2})\lambda$ *(Destructive interference)*
   d = the slit separation
4. Phase difference between two waves $= \dfrac{2\pi(|\mathbf{r}_2| - |\mathbf{r}_1|)}{\lambda}$

   where $\mathbf{r}_1, \mathbf{r}_2$ are distances from wave source one and wave source two

5. Single-slit diffraction:

$$I(\theta) = I_{\max}\left(\frac{\sin\alpha}{\alpha}\right)^2 \qquad \alpha = \frac{\pi a \sin\theta}{\lambda};$$

where a = slit width

6. N slit diffraction:

$$I(\theta) = N^2 I_0 \left(\frac{\sin^2\alpha}{\alpha^2}\right)\left(\frac{\sin^2 N\beta}{\sin^2\beta}\right)$$

where α, β = as above

N = number of equally spaced identical slits

Program 11.2 (a)

NSLITBAS *N slit diffraction pattern (normalized to 1)*

```
100 LET L0=1
110 LET X0=.005
120 PRINT "SLIT-SCREEN DIST.=";L0;" M.; SCREEN WIDTH=";1E3*X0;"MM."
130 PRINT
140 PRINT "# OF SLITS (ZERO TO FINISH)";
150 INPUT N
160 IF N=0 THEN 500
170 PRINT "SLIT WIDTH (MILLIMETERS)";
180 INPUT A
190 LET A=A*1E-3
200 IF N=1 THEN 240
210 PRINT "SLIT SEPARATION (MILLIMETERS)";
220 INPUT H
230 LET H=H*1E-3
240 PRINT "WAVELENGTH (ANGSTROMS)";
250 INPUT L
260 LET L=L*1E-10
270 PRINT "POSITION","INTENSITY"
280 LET X5=X0/(32*N)   Interval for step across screen
290 FOR X1=-X0/2 TO X0/2 STEP X5   Step across screen
300 LET S=X1/L0   Sine (θ) for small θ
310 IF S<>0 THEN 350  }
320 LET Y1=1          } Center of
330 PRINT X1,Y1       } pattern
340 GOTO 470          }
350 LET A1=3.14159*A*S/L   π width sin(θ)/λ ≡ α
360 IF N>1 THEN 400
370 LET S5=SIN(A1)  }
380 LET S4=1        } Single
385 LET S6=1        } slit
390 GOTO 450        }
400 LET D=3.14159*H*S/L   π separation sin(θ)/λ ≡ β
410 LET S4=SIN(D)   sin(β)
420 IF S4=0 THEN 320
430 LET S5=SIN(A1)   sin(α)
440 LET S6=SIN(N*D)   sin(Nβ)
450 LET Y1=S5*S5*S6*S6/(A1*A1*N*N*S4*S4)   N slit formula
460 PRINT X1,Y1
470 NEXT X1
480 PRINT
490 GOTO 140   Return for new geometry
500 END
```

Program 11.2 (b)

```
NSLITFOR      N slit diffraction pattern (normalized to 1)

          TOSCRN=1.   Distance to screen
          XWIDTH=.001    Width of screen
  10      READ(5,100)SLTNUM,SLTWID,SLTSEP,WAVEL
          IF (SLTNUM.EQ.0.) GOTO 40
          WAVEL=1.E-10*WAVEL     Wavelength, λ
          DO 20 I=1,21,1     Step across screen
          AI=I-1
          X=-XWIDTH/2.+AI*XWIDTH/20.
          SINT=X/TOSCRN   sin(θ) for small θ
          ALPHA=3.14159*SLTWID*SINT/WAVEL      π width sin(θ)/λ ≡ α
          SINALP=SIN(ALPHA)
          BETA=3.14159*SLTSEP*SINT/WAVEL      π separation sin(θ)/λ ≡ β
          SINBET=SIN(BETA)
          IF (SINBET.NE.0.) GOTO 30  }
          Y=1.                       } Center
          GOTO 20                    } of pattern
  30      SINNBE=SIN(SLTNUM*SLTSEP)   sin(Nβ)
          Y=SINALP**2*SINNBE**2/(ALPHA**2*SLTNUM**2*SINBET**2)   N slit formula
  20      PRINT 101,X,Y
          GOTO 10   Return for new geometry
  100     FORMAT(4F5.4)
  101     FORMAT(1X,2E15.5)
  40      STOP
          END
```

11.6 Programs

In the program NSLIT [Programs 11.2(a), (b)], the computer calculates the closed-form solutions to the N-slit diffraction problem. IVSANG calculates the intensity pattern as a function of the angle around a general distribution of wave sources. THNFLM calculates the interference effects in a thin film of dielectric material (oil on glass).

For the programs AMPLI [Programs 11.3(a), (b)], INTEN [Programs 11.4(a), (b)], and IVSANG [Programs 11.5(a), (b), (c)] you may place wave sources anywhere in the xy plane. The program AMPLI calculates the amplitudes of the waves from each source reaching some specified point and then prints the total amplitude. The program INTEN calculates the intensity averaged over phase at some specified observation point. The program IVSANG prints intensities at a number of points around a large circle. After calculating the sum of the wave amplitudes for a number of different times (or phases), the program prints the squares of the amplitudes. The result is the intensity (due to all the sources) at each point around a large circle (the intensity as a function of angle).

Figure 11.6 shows the intensity versus angle plotted for the two-source pattern of Figure 11.5 (Young's double-slit experiment). This illustrates the use of IVSANG. The broad maxima are the ±90° (constructive interference) directions. The other lobes occur at 0 and ±30° in the forward and backward directions.

Program 11.3 (a)

AMPLIBAS *Amplitude due to wave sources in 2D*

```
100 DIM X(10),Y(10),A(10),P(10)
110 PRINT "NUMBER OF SOURCES, WAVELENGTH";
120 INPUT N,L
130 PRINT "FOR EACH SOURCE:"
140 PRINT "INPUT COORDINATES (X,Y), AMPLITUDE, AND PHASE (DEGS.)."   Geometry of wave sources
150 FOR I=1 TO N
160 INPUT X(I),Y(I),A(I),P(I)
170 LET P(I)=3.14159*P(I)/180
180 NEXT I
190 PRINT "COORDINATES OF THE OBSERVATION POINT";
200 INPUT X0,Y0
210 PRINT "AMPLITUDE OVER ONE CYCLE"
220 FOR P0=0 TO 6.28318-3.14159/8 STEP 3.14159/8   Step through phase
230 LET A1=0
240 FOR I=1 TO N   Sum over sources
250 LET X1=X0-X(I)
260 LET Y1=Y0-Y(I)                Distance from each source to
270 LET R=SQR(X1*X1+Y1*Y1)        observation point
280 LET A1=A1+A(I)*SIN(6.28318*R/L+P(I)+P0)   Amplitude sum
290 NEXT I
300 PRINT A1
310 NEXT P0
320 GOTO 190   Return for new observation point
330 END
```

Program 11.3 (b)

AMPLIFOR *Amplitude due to several wave sources in 2D*

```
      DIMENSION X(10),Y(10),A(10),P(10)
      DATA N,AL/4,1./
      DATA Y(1),Y(2),Y(3),Y(4)/4*0./
      DATA X(1),X(2),X(3),X(4)/-1.5,-.5,.5,1.5/      Geometry of wave sources
      DATA A(1),A(2),A(3),A(4)/4*1./
      DATA P(1),P(2),P(3),P(4)/4*0./
5     READ(5,100)XOBS,YOBS   Observation point
      IF (XOBS-999.) 7,30,7
7     DO 20 J=1,16   Step through 2π in phase
      PHASE=(J-1)*3.14159/8.
      AMPL=0.
      DO 10 I=1,N   Sum over sources
      DX=XOBS-X(I)
      DY=YOBS-Y(I)                 Distance from each source to
      R=SQRT(DX*DX+DY*DY)          observation point
10    AMPL=AMPL+A(I)*SIN(6.28318*R/AL+P(I)+PHASE)   Amplitude sum
20    PRINT 101,AMPL
      GOTO 5
30    STOP
100   FORMAT(2F10.4)
101   FORMAT(1X,F10.4)
      END
```

Program 11.4 (a)

INTENBAS *Intensity due to several wave sources*

```
100 DIM X(10),Y(10),A(10),P(10)
110 PRINT "NUMBER OF SOURCES, WAVELENGTH";
120 INPUT N,L
130 LET N9=0
140 PRINT "FOR EACH SOURCE:"
150 PRINT "INPUT COORDINATES (X,Y), AMPLITUDE, AND PHASE (DEGS.)."   Geometry of wave sources
160 FOR I=1 TO N
170 INPUT X(I),Y(I),A(I),P(I)
180 LET P(I)=3.14159*P(I)/180
190 LET N9=N9+A(I)
200 NEXT I
210 LET N9=N9*N9   Normalization constant
220 PRINT "COORDINATES OF THE OBSERVATION POINT";
230 INPUT X0,Y0
240 LET I0=0
250 FOR P0=0 TO 6.28318-3.14159/8 STEP 3.14159/8   Sum over one cycle
260 LET A1=0
270 FOR I=1 TO N   Sum over sources
280 LET X1=X0-X(I)            Distance from
290 LET Y1=Y0-Y(I)            each source to
300 LET R=SQR(X1*X1+Y1*Y1)    observation point
310 LET A1=A1+A(I)*SIN(6.28318*R/L+P(I)+P0)   Amplitude sum
320 NEXT I
330 LET I0=I0+A1*A1   Intensity
340 NEXT P0
350 PRINT "INTENSITY =";I0/(8*N9)
360 GOTO 220
370 END
```

Program 11.4 (b)

INTENFOR *Intensity due to several wave sources*

```
      DIMENSION XSOURC(10),YSOURC(10),AMPL(10),PHASE(10)
      READ(5,100)NSOURC,WAVEL
      AMPNOR=0.
      DO 10 I=1,NSOURC,1
      READ(5,101)XSOURC(I),YSOURC(I),AMPL(I),PHASE(I)   Geometry of wave sources
      PHASE(I)=3.14159*PHASE(I)/180.
10    AMPNOR=AMPNOR+AMPL(I)
      AMPNOR=AMPNOR**2   Normalization constant
20    READ(5,102)XOBS,YOBS   Observation constant
      IF (XOBS.EQ.9999.) GOTO 50
      AINTEN=0.
      DO 40 IP=1,16,1   Sum over one cycle
      AIP=IP-1
      AVPHAS=AIP*3.14159/8.
      AVAMP=0.
      DO 30 IN=1,NSOURC,1   Sum over sources
      X=XOBS-XSOURC(IN)     Distance from
      Y=YOBS-YSOURC(IN)     each source to
      R=SQRT(X**2+Y**2)     observation point
30    AVAMP=AVAMP+AMPL(IN)*SIN(6.28318*R/WAVEL+PHASE(IN)+AVPHAS)   Amplitude sum
40    AINTEN=AINTEN+AVAMP**2/(8.*AMPNOR)   Intensity
      PRINT 103,XOBS,YOBS,AINTEN
      GOTO 20
100   FORMAT(I5,F5.2)
101   FORMAT(4F5.2)
102   FORMAT(2F5.2)
103   FORMAT(1X,3E15.5)
50    STOP
      END
```

Program 11.5 (a) Block diagram for the strategy to calculate polar-intensity plots for arbitrary sets of wave sources.

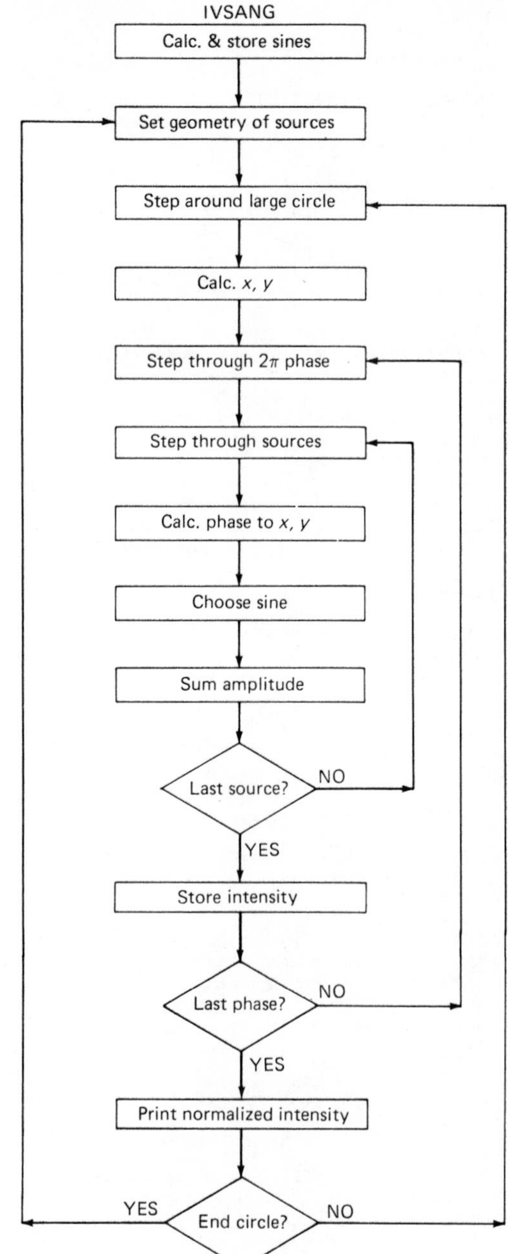

Program 11.5 (b)

IVSANGBA *Intensity versus angle for several sources (2D)*

```
100 DIM I(200),S(361),X(10),Y(10),A(10),P(10)
110 FOR I=1 TO 361
120 LET T0=(I-1)*6.28318/360          Store sines
130 LET S(I)=SIN(T0)                  needed later
140 NEXT I
150 PRINT "HOW MANY SOURCES, WHAT WAVELENGTH";
160 INPUT N9,L0
170 LET N8=N9*N9
180 PRINT "(X,Y), AMPL. AND PHASE (DEGS.) FOR EACH SOURCE:"   Geometry
190 FOR I=1 TO N9                                              of wave
200 PRINT "SOURCE";I;                                          sources
210 INPUT X(I),Y(I),A(I),P
220 LET P(I)=3.14159*P/180
230 NEXT I
240 PRINT "HOW MANY POINTS AROUND OBSERVATION CIRCLE";
250 INPUT N7
260 LET J0=0
270 LET R0=10    Radius of observation circle
280 FOR P0=0 TO 6.28318 STEP 6.28318/N7    Step around circle
290 LET J0=J0+1
300 LET X0=R0*COS(P0)    Coordinates of each
310 LET Y0=R0*SIN(P0)    point on circle
320 LET N5=8
330 FOR P5=0 TO 6.28318-3.14159/N5 STEP 3.14159/N5    Average over phase
340 LET I9=0
350 FOR I=1 TO N9    Sum over sources
360 LET X1=X0-X(I)                r from each
370 LET Y1=Y0-Y(I)                source to point
380 LET R1=SQR(X1*X1+Y1*Y1)       on circle
390 LET P4=2*3.14159*R1/L0+P(I)   Phase of each
400 LET P=P4+P5                   wave at circle
410 LET P=P-INT(P/(2*3.14159))*(2*3.14159)    Choose right
420 LET J1=INT(P/(3.14159/180))               stored sine
430 LET I9=I9+A(I)*S(J1)    Amplitude sum
440 NEXT I
450 LET I(J0)=I(J0)+I9*I9/N8    Intensity average
460 NEXT P5
470 NEXT P0
480 PRINT "POINT","ANGLE","INTENSITY"
490 FOR J=1 TO J0
500 LET I9=I(J)                                     Print
510 LET I9=I9/N5                                    group
520 PRINT J,(J-1)*(6.28318/N7)*180/3.14159,I9
530 NEXT J
540 PRINT
550 GOTO 150    Return for new source
560 END
```

Program 11.5 (c)

```
IVSANGFO      Intensity versus angle for several sources (2D)

          DIMENSION AIN(200),S(361),X(10),Y(10),AMP(10),PHASE(10)
          DO 10 I=1,361                  } Store sines
          T0=(I-1)*6.28318/360.          } needed later
10        S(I)=SIN(T0)
          DATA NSOUR,AL0/2,1./   Number of sources, wavelength      } Geometry
          ANCIR=NSOUR*NSOUR                                          } of wave
          DATA X(1),Y(1),AMP(1),PHASE(1)/.5,0.,1.,0./                } sources
          DATA X(2),Y(2),AMP(2),PHASE(2)/-.5,0.,1.,0./
          DATA NCIRC/16/   Number of points around circle
          ANCIRC=NCIRC
          R0=10.   Radius of observation circle
          DO 80 J0=1,NCIRC+1   Step around circle
          ANGC=(J0-1)*6.28318/ANCIRC
          X0=R0*COS(ANGC)     } Coordinates of each
          Y0=R0*SIN(ANGC)     } point on circle
          NAV=16
          ANAV=NAV
          DO 80 IP=1,NAV   Average over phase
          PHAAV=(IP-1)*6.28318/ANAV
          AMPLI=0.
          DO 70 I=1,NSOUR   Sum over sources
          XOBS=X0-X(I)                       } r from each
          YOBS=Y0-Y(I)                       } source to point
          ROBS=SQRT(XOBS**2+YOBS**2)         } on circle
          PHA=6.28318*ROBS/AL0+PHASE(I)   } Phase of each
          REPHA=PHA+PHAAV                 } wave at circle
          REPHA=REPHA-AINT(REPHA/(6.28318))*(6.28318)   } Choose right
          J1=1+AINT(REPHA/(3.14159/180.))               } stored sine
70        AMPLI=AMPLI+AMP(I)*S(J1)   Amplitude sum
80        AIN(J0)=AIN(J0)+AMPLI*AMPLI/ANCIR   Intensity average
          DO 85 J=1,NCIRC+1                              }
          AI=2.*AIN(J)/ANAV                              } Print
          ANG=(J-1)*(6.28318/ANCIRC)*180./3.14159        } group
85        PRINT 100,J,ANG,AI                             }
          STOP
100       FORMAT(1X,I5,2F10.4)
          END
```

One problem which can be treated using the program IVSANG concerns antenna arrays. Have you ever driven by a radio station and seen what appear to be several, identical, regularly spaced antennas? Often the plane formed by the antennas faces a nearby city or urban area. Many of these arrays are broadside arrays. By using the interference pattern from several sources, broadside arrays can direct most of their signal in one direction. The city gets better reception and receivers on either side of the array get nothing at all. You can calculate the effects of broadside-array interference using IVSANG (or your own version of the program). Antenna design is quite advanced, even though it is largely a black art. See H. P. Westman (ed.), *Reference Data for Radio Engineers* (Howard W. Sams, Indianapolis, 1970), 5th ed., and the references given therein for more information on antenna arrays. You can try to find directional patterns for different arrays of antennas with IVSANG by placing sources in two-dimensional arrays.

Figure 11.7(a) shows the intensity-versus-angle plot for a broadside array of four

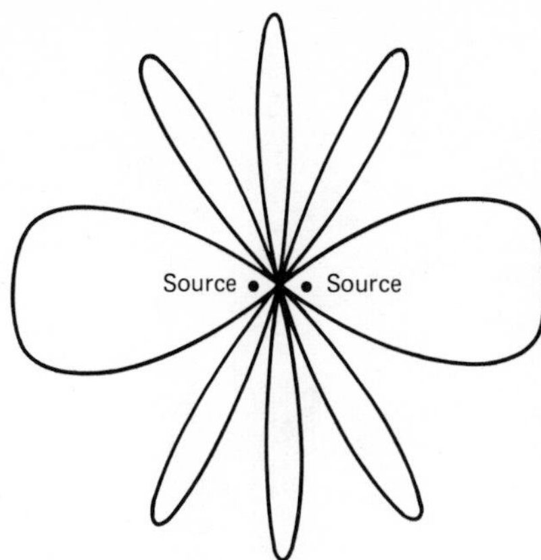

Figure 11.6 The polar plot of intensity versus angle for the geometry of Figure 11.5. This diagram shows the full 360° pattern. Notice the wider lobes in the sideward directions and that the angular positions of the lobes agree with the maxima in Figure 11.5. The program allows sources of arbitrary amplitude and phase to be placed anywhere in the plane. The program then calculates the intensities on a large circle surrounding all the sources.

antennas placed $\frac{1}{2}\lambda$ apart and fed equal in-phase currents. Most of the intensity is in a narrow range of angles in the forward (and backward) directions. Figure 11.7(b) shows the pattern from a four-antenna, binomial broadside array. Here the antennas are fed different (although still in-phase) currents. There are now no side lobes of appreciable intensity and the breadth of the high-intensity lobes is somewhat wider. What happens if you feed the antennas currents of different phases? Can you get rid of the backward lobe by using some suitable two-dimensional array of antenna positions?

Figure 11.8 is a display of (normalized) intensity versus position on a screen for one, two, and five slits (or sources). This illustrates the use of the program NSLIT. The $N = 2$ pattern is what you would find by placing a flat screen parallel to the line between the sources of Figure 11.5 or 11.6. This projection-screen geometry is the one almost always used in laboratory demonstrations of Young's double-slit experiment. After studying Figure 11.8, see if you can sketch the polar intensity versus angle for five slits. See if you can decide what the intensity patterns on screens would look like if the screens were put at various places and at various orientations to the sources for Figure 11.6.

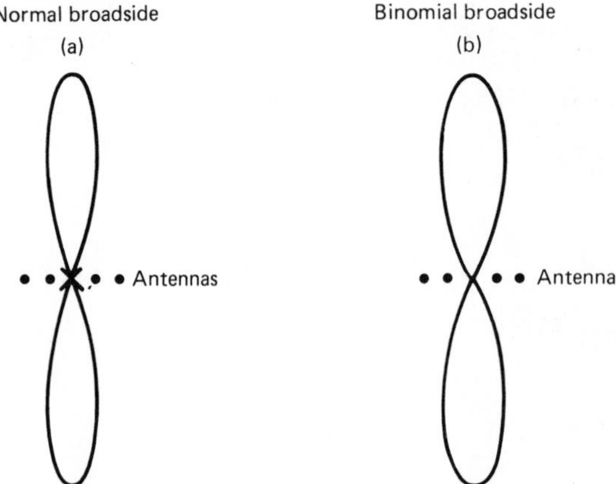

Figure 11.7 Intensity-versus-angle patterns for four-element radio-antenna-broadside arrays. The antennas are placed half-wavelengths apart and so have maximum transmission in the forward and backward directions. Curve (a) shows the pattern with equal in-phase currents fed to all four antennas. Notice the small side lobes. Curve (b) shows the pattern for the so-called binomial array. The currents fed to the antennas are in phase but have different amplitudes.

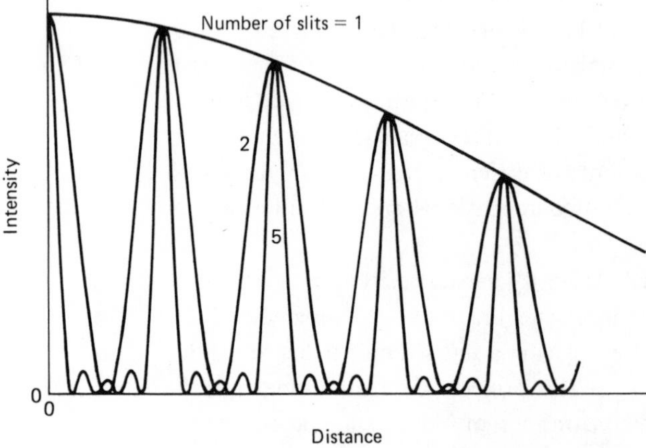

Figure 11.8 Intensity pattern on the screen of an N-slit diffraction pattern. The intensity from one, two, and five identical slits are shown as functions of position across the screen.

11.7 Thin-Film Interference

Thin-film interference phenomena offer another application of computers. The basic phenomenon is easy to understand. A wave impinges at angle θ onto a plane-parallel film whose thickness is of the order of the wavelength. Some of the wave reflects and some refracts into the second medium. Of the part in the second medium, some reflects at the back face, and some is transmitted out the back into the third medium. This process repeats (in principle an infinite number of times), and each wave is different in amplitude and phase from its neighbors. When the waves are brought together (by a lens, for example), they interfere. Maxima and minima occur due to constructive and destructive interference as the wavelength λ, the film thickness d, or the angle of incidence θ is varied.

By representing the wave by lines perpendicular to wavefronts, the phenomenon can be drawn diagrammatically, as in Figure 11.9. Call the ratio of reflected amplitude to incident amplitude (the reflection coefficient) at an $n_1 \longrightarrow n_2$ interface R_1 and the corresponding transmission ratio (the transmission coefficient) T_1. Call the ratios at an $n_2 \longrightarrow n_1$ interface R_2 and T_2. Assume that the third index of refraction n_3 is equal to the first, n_1. Then for the ith wave on the front side, the amplitude is

$$A(I) = AT_1 (R_2)^{2i-1} T_2$$

with

$$A(0) = AR_1$$

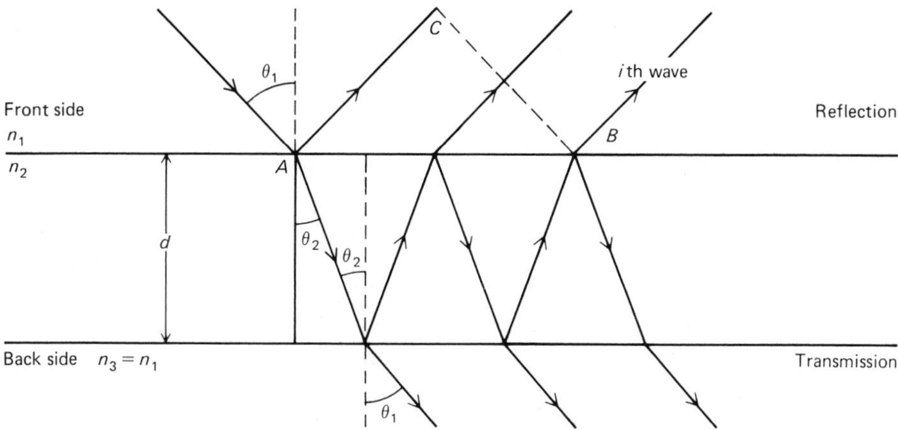

Figure 11.9 The geometry of a multiple-reflection–transmission pattern in a thin film of index of refraction n_2.

Program 11.6 (a)

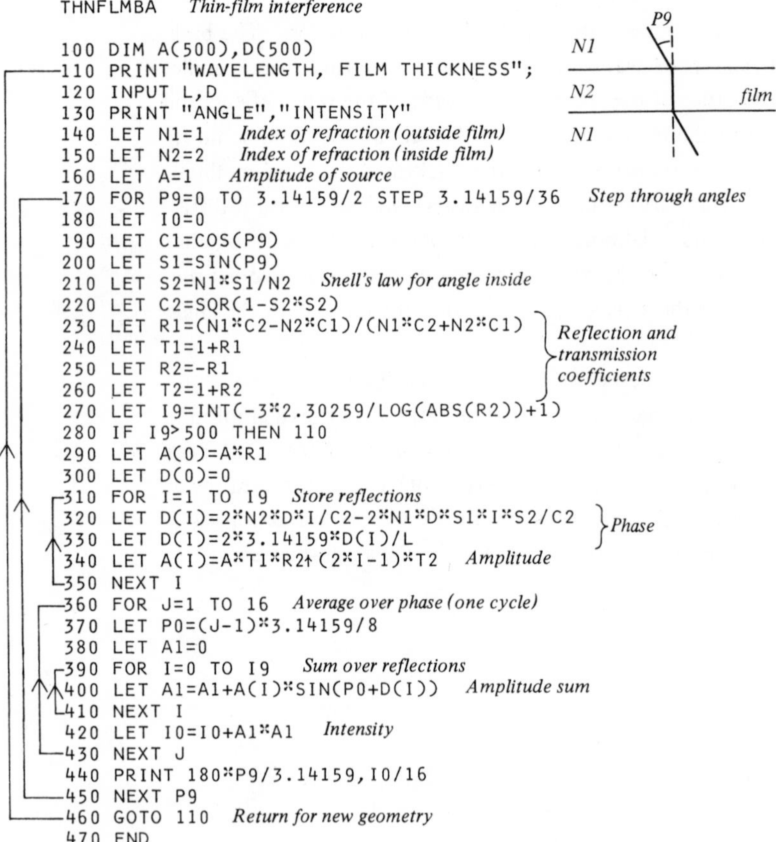

```
THNFLMBA     Thin-film interference

100 DIM A(500),D(500)
110 PRINT "WAVELENGTH, FILM THICKNESS";
120 INPUT L,D
130 PRINT "ANGLE","INTENSITY"
140 LET N1=1      Index of refraction (outside film)
150 LET N2=2      Index of refraction (inside film)
160 LET A=1     Amplitude of source
170 FOR P9=0 TO 3.14159/2 STEP 3.14159/36    Step through angles
180 LET I0=0
190 LET C1=COS(P9)
200 LET S1=SIN(P9)
210 LET S2=N1*S1/N2     Snell's law for angle inside
220 LET C2=SQR(1-S2*S2)
230 LET R1=(N1*C2-N2*C1)/(N1*C2+N2*C1)  } Reflection and
240 LET T1=1+R1                         } transmission
250 LET R2=-R1                          } coefficients
260 LET T2=1+R2
270 LET I9=INT(-3*2.30259/LOG(ABS(R2))+1)
280 IF I9>500 THEN 110
290 LET A(0)=A*R1
300 LET D(0)=0
310 FOR I=1 TO I9   Store reflections
320 LET D(I)=2*N2*D*I/C2-2*N1*D*S1*I*S2/C2  } Phase
330 LET D(I)=2*3.14159*D(I)/L               }
340 LET A(I)=A*T1*R2↑(2*I-1)*T2    Amplitude
350 NEXT I
360 FOR J=1 TO 16   Average over phase (one cycle)
370 LET P0=(J-1)*3.14159/8
380 LET A1=0
390 FOR I=0 TO I9     Sum over reflections
400 LET A1=A1+A(I)*SIN(P0+D(I))    Amplitude sum
410 NEXT I
420 LET I0=I0+A1*A1     Intensity
430 NEXT J
440 PRINT 180*P9/3.14159,I0/16
450 NEXT P9
460 GOTO 110   Return for new geometry
470 END
```


where A is the incident wave amplitude. For the ith wave out the back,

$$A(I) = AT_1 (R_2)^{2(i-1)} T_2$$

$$A(0) = 0$$

The interference effect also depends on the phase differences between the waves as they emerge from the film. A lens brings together all the parallel waves from points such as A and B in the sketch. While the ith ray was reaching B, it traveled an optical path length of $2n_2\, di/\cos\theta_2$. (The optical path length is the index of refraction times the true geometrical length.) The point on the first wave on the same wavefront (point C in the sketch) has traveled an optical path length of $2n_1\, di \sin\theta_1 \tan\theta_2$. The phase

Program 11.6 (b)

```
THNFLMFO    Thin-film interference

        DIMENSION A(500),D(500)
10      READ(5,100)ALAM,D    λ, thickness
        IF (ALAM-999.) 20,60,20
20      AN1=1.    Index of refraction (outside film)
        AN2=2.    Index of refraction (inside film)
        A=1.   Amplitude of source
        DO 50 K=1,19    Step through angles
        ANG=(K-1)*3.14159265/36.
        AI0=0.
        C1=COS(ANG)
        S1=SIN(ANG)
        S2=AN1*S1/AN2    Snell's law for angle inside
        C2=SQRT(1-S2*S2)
        R1=(AN1*C2-AN2*C1)/(AN1*C2+AN2*C1)   }
        T1=1.+R1                             } Reflection and
        R2=-R1                               } transmission
        T2=1.+R2                             } coefficients
        I9=IFIX(-3.*2.30259/ALOG(ABS(R2))+2)
        IF (I9-500) 20,20,10
20      A(1)=A*R1
        D(1)=0.
        DO 30 I=2,I9    Store phases and amplitudes of reflections
        D(I)=2.*AN2*D*I/C2-2.*AN1*D*S1*I*S2/C2
        D(I)=2.*3.14159265*D(I)/ALAM
30      A(I)=A*T1*R2**(2*I-1)*T2
        DO 45 J=1,16    Average over phase
        P0=(J-1)*3.14159265/8.
        A1=0.
        DO 40 I=1,I9
40      A1=A1+A(I)*SIN(P0+D(I))   Amplitude
45      AI0=(AI0+A1*A1)   Intensity
        AI0=AI0/16.
        DANG=180.*ANG/3.14159265
50      PRINT 101,DANG,AI0
        GOTO 10
60      STOP
100     FORMAT(2F10.4)
101     FORMAT(1X,2F10.4)
        END
```

difference between these waves is

$$\frac{2\pi(\text{difference in optical path lengths})}{\lambda}$$

The computer calculates these amplitudes and phase differences, and then adds the resulting waves. The programming must average over a full cycle of the incident wave in order to make sure that a zero amplitude is not just an artifact of the particular phase of the incident wave (the moment at which time was stopped). The program THNFLM [Programs 11.6(a), (b)] is one way to implement the ideas.

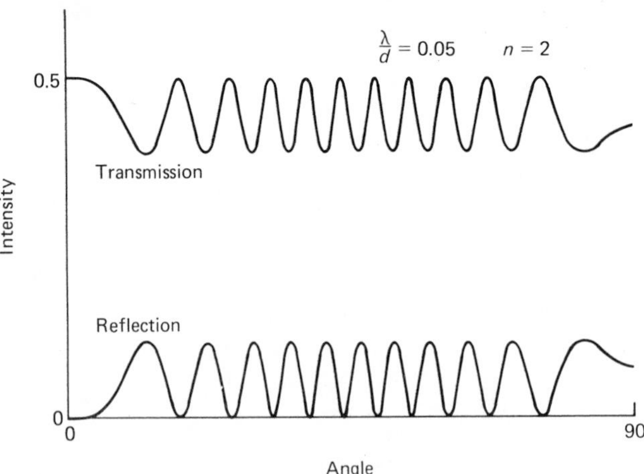

Figure 11.10 Transmitted and reflected thin-film intensities as functions of incident angle θ. The ratio of wavelength to film thickness (λ/d) is 0.05; the film has an index of refraction of 2 and is immersed in air ($n = 1$). The reflection and transmission coefficients are constant at $R_1 = 0.25$, $T_1 = 0.75$, $R_2 = 0.25$, and $T_2 = 1.25$.

Figure 11.10 shows the intensity patterns as a function of the incident angle for $\lambda/d = 0.05$, $n_1 = 1$, and $n_2 = 2$. The reflection and transmission coefficients are assumed to be constant: $R_1 = -0.25$, which means 25% of the amplitude is reflected and the wave suffers a 180° phase change in the reflection; $T_1 = 1 + R_1$; $R_2 = -R_1$; and $T_2 = 1 + R_2$. Thus in Figure 11.10, $R_1 = -0.25$, $T_1 = 0.75$, $R_2 = 0.25$, and $T_2 = 1.25$. Bright and dark fringes are observed both in the transmitted beam and the reflected beam. The maximum intensity is 0.5 for the intensity normalization chosen. The reflected beam intensity is zero for $\theta = 0$; and the transmitted beam intensity is a maximum because exactly 80 wavelengths fit down and back inside the film. The waves come back in phase with the incident wave and end up exactly out of phase with the first reflected wave.

Figure 11.11 shows the intensity versus angle for $\lambda/d = 0.05$, $n_1 = 1$, $n_2 = 2$, but $R_1 = -0.5$. Thus $T_1 = 0.5$, $R_2 = 0.5$, and $T_2 = 1.5$. Again fringes are observed, but the dark fringes are considerably deeper and sharper. This is a general result. The positions of maxima and minima are set by phase differences, but the depth and sharpness of the fringes are determined by the reflection and transmission coefficients.

These figures show the correct general behavior, but some results are clearly unphysical. For example, the amount of the wave transmitted through the film should go to zero at 90°, since $\theta = 90°$ is grazing incidence. Furthermore, for real films one should observe amplitude effects at Brewster's angle for correctly linearly polarized

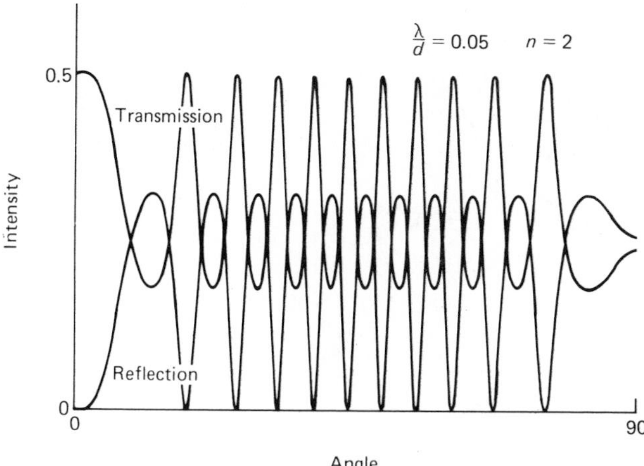

Figure 11.11 Transmitted and reflected intensities as functions of angle. $\lambda/d = 0.05$, $n = 2$ for the film, and the reflection and transmission coefficients are constant with $R_1 = -0.5$. The dark fringes are deeper and sharper than in Figure 11.9.

light, and the reflection coefficient R_2 should show total internal reflection for large angles θ_2. All these effects can be included by using the correct reflection and transmission coefficients. These correct coefficients (sometimes called Fresnel coefficients) are derived from Maxwell's equations and the proper boundary conditions for the **E** and **B** fields in the electromagnetic wave at the interface. The equations predict different coefficients for light with **E** in the plane of incidence ($E_\parallel$) than for **E** perpendicular to the plane of incidence ($E_\perp$). For $E_\parallel$, $T_1 = 1 + R_1$, $R_2 = -R_1$, and $T_2 = 1 + R_2$. For $E_\perp$, $T_1 = (\cos\theta_1)(1 + R_1)/\cos\theta_2$, $R_2 = -R_1$, and $T_2 = (\cos\theta_2)(1 + R_2)/\cos\theta_1$. So all we need to quote are the reflection coefficients R_1 for the two polarizations. The results are

For $E_\perp$:

$$R_1 = \frac{n_1 \cos\theta_1 - n_2 \cos\theta_2}{n_1 \cos\theta_1 + n_2 \cos\theta_2}$$

For $E_\parallel$:

$$R_1 = \frac{n_1 \cos\theta_2 - n_2 \cos\theta_1}{n_1 \cos\theta_2 + n_2 \cos\theta_1}$$

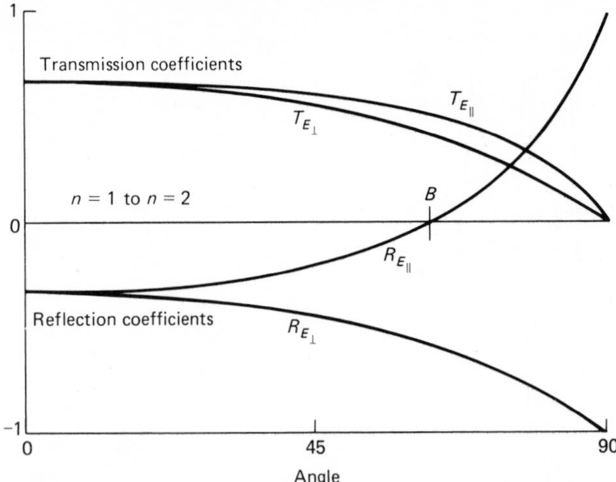

Figure 11.12 Correct reflection (R) and transmission (T) coefficients for E in the plane of incidence ($E_\parallel$) and **E** perpendicular to the plane of incidence ($E_\perp$). The wave is going from a region where $n = 1$ to a region where $n = 2$. Brewster's angle (B) is about 63.5°.

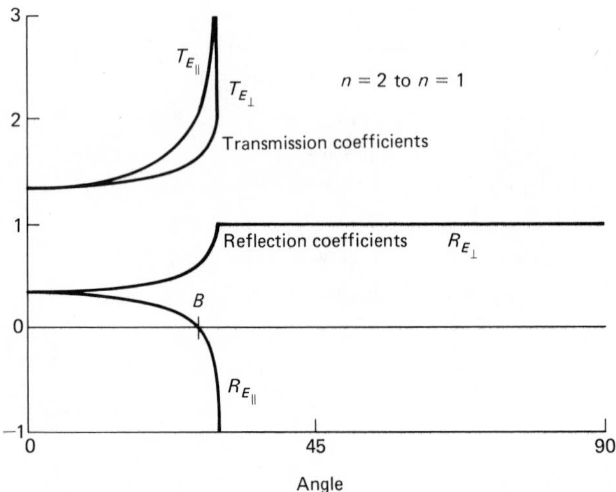

Figure 11.13 Correct transmission and reflection coefficients as functions of angle for $E_\parallel$ and $E_\perp$. The wave is at an interface between $n = 2$ and $n = 1$. Brewster's angle is at 26.6°, and total internal reflection occurs for angles greater than 30° ($\sin\theta = 0.5$).

Figure 11.12 shows the reflection and transmission coefficients through an interface between $n_1 = 1$ and $n_2 = 2$. The transmission coefficient T goes to zero as θ reaches 90° for both polarizations. Brewster's angle, marked B, occurs at about 63.5° for $E_{\parallel}$.

Figure 11.13 shows reflection and transmission coefficients for an $n_1 = 2$ to $n_2 = 1$ interface. Brewster's angle is at 26.6°, and total internal reflection occurs for angles greater than 30° ($\sin\theta = 0.5$). These figures are plotted from output of a program such as REFCOE [Programs 11.7(a), (b)].

Program 11.7 (a)

REFCOEBA *Fresnel reflection and transmission coefficients*

```
100 PRINT "N1,N2(>N1),THETA (DEGS.)";
110 INPUT N1,N2,T0
120 LET T0=3.14159*T0/180
130 LET S1=SIN(T0)
140 LET C1=COS(T0)
150 LET S2=N1*S1/N2        Snell's law; sine in second medium
160 LET C2=SQR(1-S2*S2)    Cosine in second medium
170 PRINT "ANGLE IN 2ND MEDIUM =";180*ATN(S2/C2)/3.14159
180 LET R1=(N1*C2-N2*C1)/(N1*C2+N2*C1)   p reflection  } Rare to
190 LET S1=(N1*C1-N2*C2)/(N1*C1+N2*C2)   s reflection  } dense
200 LET T1=C1*(1+R1)/C2    p transmission              } medium
210 LET U1=1+S1     s transmission                     }
220 PRINT "FOR 1-2: RP,RS,TP,TS"
230 PRINT R1,S1,T1,U1
240 LET R2=-R1    p reflection                 } Dense
250 LET S2=-S1    s reflection                 } to rare
260 LET T2=C2*(1+R2)/C1    p transmission      } medium
270 LET U2=1+S2    s transmission              }
280 PRINT "FOR 2-1: RP,RS,TP,TS"
290 PRINT R2,S2,T2,U2
300 PRINT
310 GOTO 100    Return for new geometry
320 END
```

Program 11.7 (b)

REFCOEFO *Fresnel reflection and transmission coefficients*

```
5       READ(5,100)AN1,AN2,THETA
        IF (AN1-999.) 10,20,20
10      THETA=3.14159*THETA/180.
        SN1=SIN(THETA)
        CS1=COS(THETA)
        SN2=AN1*S1/AN2    Snell's law; sine in second medium
        CS2=SQRT(1-SN2*SN2)    Cosine in second medium
        RP1=(AN1*CS2-AN2*CS1)/(AN1*CS2+AN2*CS1)   p reflection  } Rare to
        RS1=(AN1*CS1-AN2*CS2)/(AN1*CS1+AN2*CS2)   s reflection  } dense
        TP1=CS1*(1.+RS1)/CS2    p transmission                  } medium
        TS1=1.+SN1    s transmission                            }
        PRINT 101,RP1,RS1,TP1,TS1
        RP2=-RP1    p reflection                  } Dense
        RS2=-RS1    s reflection                  } to rare
        TP2=CS2*(1.+RP2)/CS1    p transmission    } medium
        TS2=1.+SN2    s transmission              }
        PRINT 101,RP2,RS2,TP2,TS2
        GOTO 5    Return for new geometry
20      STOP
100     FORMAT(3F10.4)
101     FORMAT(1X,4F10.4)
        END
```

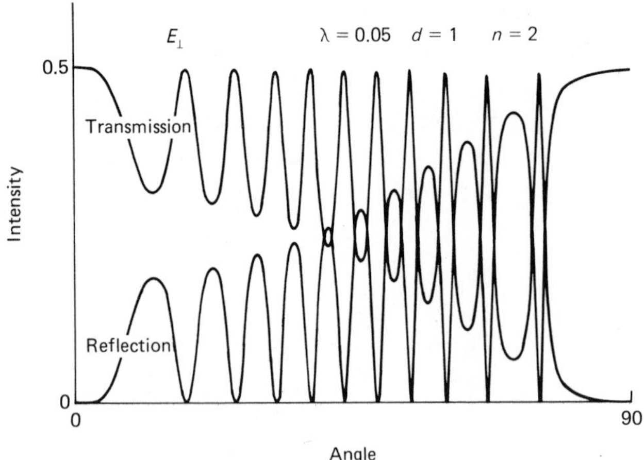

Figure 11.14 Reflected and transmitted thin-film intensities as functions of incident angle using the correct reflection and transmission coefficients $\lambda/d = 0.05$ and $n = 2$. **E** is perpendicular to the plane of incidence, so Brewster's angle does not occur. The amplitudes of the fringes show the angular dependence of the reflection and transmission coefficient.

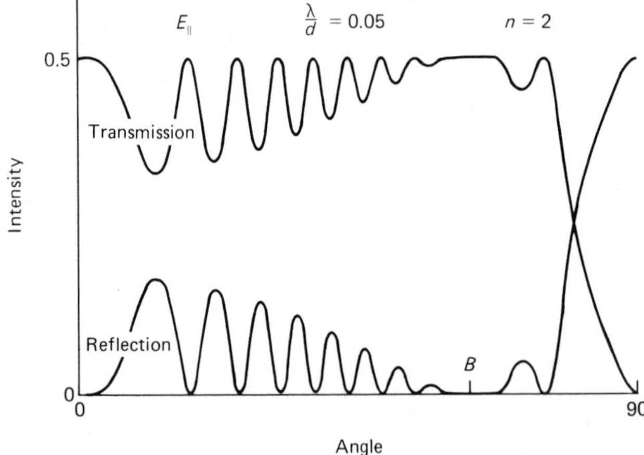

Figure 11.15 Reflected and transmitted intensities as functions of incident angle for **E** parallel to the plane of incidence. The effects of Brewster's angle B are shown.

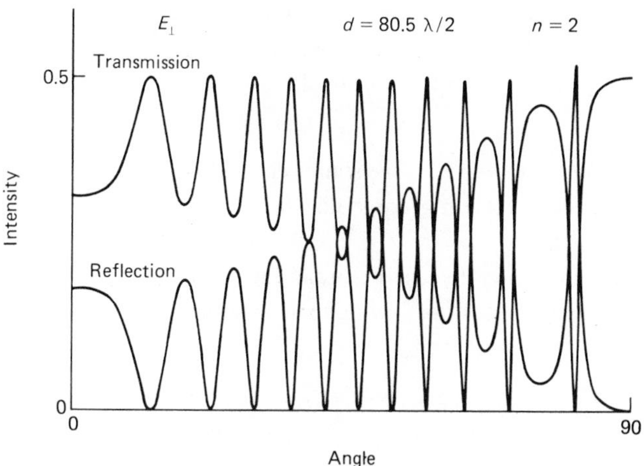

Figure 11.16 Transmitted and reflected beam intensities for $\mathbf{E}_\perp$ when 80.5 wavelengths fit in the sample at $\theta = 0$. The central reflection is now bright, and the central transmission is now dark. Otherwise the parameters are those of Figure 11.13.

If we use these coefficients in the program THNFLM, we get good physical results. Figure 11.14 shows the transmitted intensity T and reflected intensity R as functions of incident angle θ for $\lambda/d = 0.05$, $n_1 = 1$, $n_2 = 2$, and $E_\perp$. Fringes occur, and the transmission goes to zero at glancing incidence. Figure 11.15 shows transmission and reflection versus angle for $\lambda/d = 0.05$, $n_1 = 1$, $n_2 = 2$, and $E_\parallel$. Here we see the effects of Brewster's angle B.

Figure 11.16 shows transmission and reflection much as in Figure 11.14, but now the wavelength is such that $80\frac{1}{2}$ wavelengths fit inside the film at $\theta = 0$. The central reflection is "bright," and the central transmission is "dark." All the patterns have fringe amplitudes influenced by the angular dependence of the reflection and transmission coefficients.

These calculations are very complicated analytically. They are straightforward on the computer. Waves are added up while keeping track of amplitudes and phases as the waves bounce.

Problems

In this material, most problems can lead to projects. The following are a set of problems which other students have performed successfully.

1. Write a program to calculate the intensity pattern of an N source experiment. Show that as N increases (for identical slits separated the same

amounts), the sharpness of the bright lines increases and less intense lines appear. Show that the number of major fringes inside the first single-slit diffraction maximum depends on the ratio of slit separation to slit size. Demonstrate the pattern dependence on wavelength.

2. Write a program to add up waves from a number of wave sources placed anywhere in a plane. Calculate the intensity pattern at large distances. Check your calculations by plotting the intensity pattern for the double-source problem. Use your program to demonstrate the use of broadside-antenna arrays. See if you can set up a highly directional two-dimensional array having only a forward lobe.

3. Television antennas are often Yagi-Uda antenna arrays. Using the book *Reference Data for Radio Engineers* (or any better reference), find out what a Yagi-Uda array is. Discuss physically why such arrays are useful antennas. Develop a strategy on the computer to calculate the radiation pattern from such an array. Since the radiation and reception characteristics are closely related, you will have calculated the directionality of reception from a TV antenna.

12 quantum mechanics

12.1 Introduction

On a microscopic scale, all particles have wavelike behavior. The electron in a hydrogen atom would radiate all its energy and collapse into the nucleus in a microsecond if it were not in a standing-wave state around the proton. Neutrons traveling through a crystalline solid behave just like waves as they interact with the solid.

These particle waves (called wave functions or probability amplitude waves) obey a wave equation. Just as waves on strings or ripples on water obey well-known differential wave equations relating spatial coordinates and time coordinates, so also do particle waves obey the Schrödinger equation. The full three-dimensional Schrödinger equation including time is beyond the scope of this work, but you can gain valuable experience with quantum mechanical ideas by solving the simpler one-dimensional Schrödinger equation. This simpler equation corresponds to differential equations for mechanical waves. A great deal of intuition into the quantum mechanical behavior of particle waves can be obtained from a study of one-dimensional systems.

Using the computer you can study the finite square well and the one-dimensional harmonic oscillator as well as a number of problems which have no analytic closed-form solutions. This computer chapter will emphasize symmetric one-dimensional potential cases of the Schrödinger equation, that is, those for which $V(-x) = V(x)$. The chapter will also discuss examples of symmetric three-dimensional potentials and show you how to solve many such problems. As written, the programs calculate unnormalized wave functions. To normalize any eigenfunction, calculate $I = \int_{-\infty}^{\infty} \psi^2\, dx$

and then multiply your wave function ψ by $1/\sqrt{I}$. The three-dimensional method will work for any potential which is spherically symmetric $[V(\mathbf{r}) = V(|\mathbf{r}|)]$, and it can often by applied to an even broader range of situations. This chapter deals mostly with stationary-state problems, that is, with finding eigenfunctions and eigenvalues for the Schrödinger equation. One-dimensional wave-packet propagation (using the time-dependent Schrödinger equation) will be mentioned.

12.2 Basic Equations

The following basic facts are discussed and derived in your text. If something looks new, review your textbook.

1. $\dfrac{\hbar^2}{2m}\dfrac{d^2\psi}{dx^2} + V(x)\psi = E\psi$ *(Schrödinger equation for one-dimensional stationary states)*

   This equatión will be simplified by measuring V and E in Hartree atomic units ($\hbar = m = e = 1$), so that the equation you will deal with is

   $$\frac{d^2\psi}{dx^2} = 2[V(x) - E]\,\psi$$

2. For a normalized wave function (that is, one for which $\int_{-\infty}^{\infty} |\psi|^2\, dx = 1$), the square of the wave function, $|\psi(x)|^2$, is the probability density of finding the particle near the point x.

3. Boundary conditions:
   a. ψ and $d\psi/dx$ must be continuous wherever $V(x)$ is finite.
   b. $\psi \longrightarrow 0$ as $|x| \longrightarrow \infty$ [and wherever $|V(x)| \longrightarrow \infty$] for bound states.

4. If $V(-x) = V(x)$ (symmetric potential), then ψ is either completely odd $[\psi(-x) = -\psi(x)$ for all $x]$ or ψ is completely even $[\psi(-x) = \psi(x)$ for all $x]$.

5. For the infinite square well $[V(x) = 0,\ |x| < a;\ V(x) = \infty,\ |x| > a]$, there is a discrete set of stationary states with energies (eigenvalues):

   $$E_n = \left(\frac{\hbar^2}{2m}\right)\left(\frac{n^2\pi^2}{4a^2}\right) \qquad n = 1, 2, 3, \ldots$$

   and wave functions (eigenfunctions):

   $$\psi_n(x) = \text{constant} \times \sin\left(\frac{n\pi}{2a}x\right) \qquad n \text{ even}$$

   $$\psi_n(x) = \text{constant} \times \cos\left(\frac{n\pi}{2a}x\right) \qquad n \text{ odd}$$

where the constant normalizes the wave functions. (These are analagous to the standing modes of a string of length $2a$.)

6. For the one-dimensional harmonic oscillator the energies are

$$E_n = (n - \tfrac{1}{2})\hbar\omega \qquad n = 1, 2, 3, \ldots$$

where

$$\omega = \sqrt{\frac{k}{m}} \quad \text{and} \quad V(x) = \tfrac{1}{2}kx^2$$

7. In general, the wave function ψ oscillates (often nonsinusoidally) in regions where $E > V(x)$ and decays toward zero amplitude in regions where $E < V(x)$. If $V(x) = $ a constant and $E < V$, the decay toward zero is exponential $[\psi(x) \propto e^{-ax}]$. Many potentials approach this situation asymptotically as x becomes large.

8. Zero point motion: The lowest energy stationary state of a system has its energy greater than the minimum of the potential.

12.3 Method for Symmetric One-Dimensional Potentials

One easy method of solving the one-dimensional stationary-state Schrödinger equation for symmetric potentials is by starting at the center of the region and integrating out to large x in a stepwise way. Since the potential is symmetric, the wave functions are of definite parity [i.e., wholly even, $\psi(x) = \psi(-x)$, or wholly odd, $\psi(-x) = -\psi(x)$]. If the wave function is even, then (putting $x = 0$ at the symmetry point of the potential) $\psi(0) \neq 0$ and $d\psi/dx = \psi' = 0$. You can arbitrarily choose the value of $\psi(0)$; your wave function will not be normalized. Use Schrödinger's equation to calculate a new ψ' and then a new ψ for each step Δx. The method is very similar to $\mathbf{F} = ma$ calculations (Chapter 1) and is called a shooting method.

1. Choose an energy E and a value of $\psi(0)$. Set $\psi'(0)$.
2. Use Schrödinger's equation to find ψ'' at the present point.
3. Find the new ψ' as $\psi'_{\text{old}} + \psi'' \, \Delta x$ and the new ψ as $\psi_{\text{old}} + \psi' \, \Delta x$. (Using a half-stepped ψ' in the interval is often helpful.)
4. Update x to $x_{\text{old}} + \Delta x$; check to see if you are far enough out in x to stop calculating. (If you are not far enough, return to step 2.)
5. Look at ψ (far out). If your chosen E was an exact eigenvalue, ψ will go to zero. Otherwise ψ will diverge.

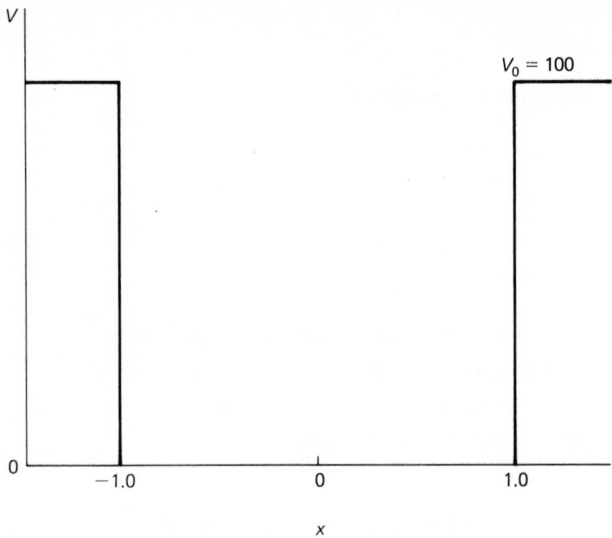

Figure 12.1 Finite-well potential. The well depth is 100; the well width is 2.

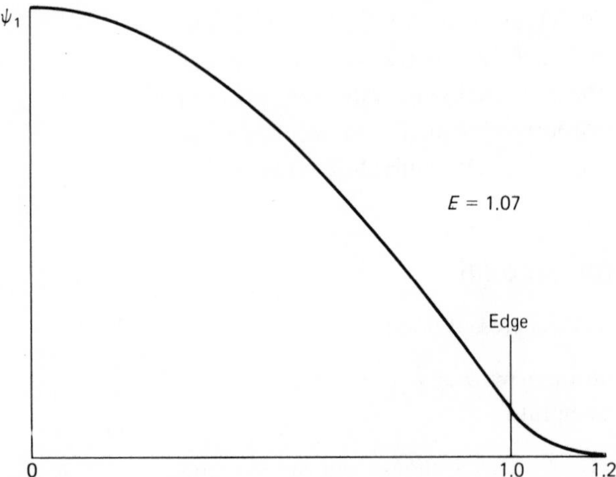

Figure 12.2 The ground-state wave function and energy for the finite well of Figure 12.1. Half the symmetric wave function is shown. The edge of the well is indicated. The energy is measured in units with $\hbar = m = e = 1$.

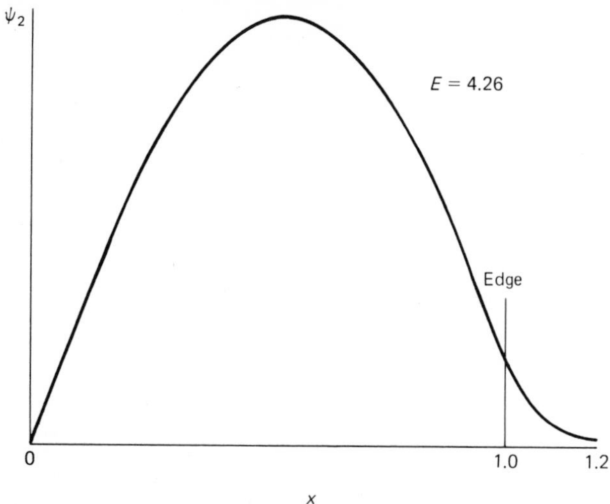

Figure 12.3 The first excited state wave function and energy for the finite well. The edge of the well is indicated. The wave function has odd symmetry.

Of course, since computers keep only finite accuracy, the wave function always diverges, but it diverges to opposite signed infinities ($+\infty$ or $-\infty$) as you pass through an eigenvalue for E. You can get four-figure accuracy for the energies quite quickly. The wave function ψ, in regions of interest, does not change much with energy; all that changes is the tail far away. Three or four pairs of energies usually define the eigenvalue to four-figure accuracy and give you a correspondingly accurate wave function.

The ground state ($n = 1$) and the odd-numbered excited states ($n = 3, 5, 7, \ldots$) are almost always even symmetry states and are solved as shown above. The even-numbered states ($n = 2, 4, 6, \ldots$) are usually of odd symmetry. For odd states, you set the wave function at the center, $\psi(0)$, equal to zero and choose an arbitrary nonzero value of $\psi'(0)$. The rest of the strategy is identical to the above.

The figures show a few examples of the method. Figure 12.1 is a picture of (one-half of) a finite square-well potential. The potential is symmetric. Figures 12.2 and 12.3 show the first two eigenfunctions (with their eigenvalues E_n). The ground state is wholly even [$\psi_1(-x) = \psi_1(x)$ for all x]; the first excited state is odd. Higher states alternate in parity. Since the well is quite deep, that is, since $V_0 >> E_1$ or E_2, the first few states approximate those of an infinite well. However, the eigenfunctions for the finite well tail off into the classically forbidden region [where $E < V(x)$], and so the eigenvalues are a little lower in energy. {The kinetic energy term in the Schrödinger equation $[(\hbar^2/2m)(d^2\psi/dx^2)]$, which depends on the curvature of the wave function, is slightly lower.}

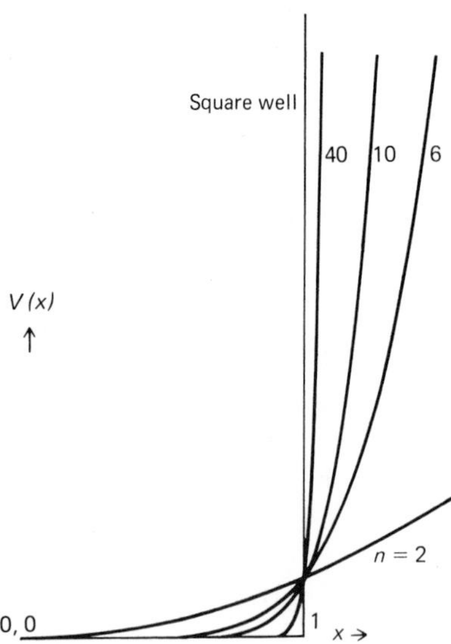

Figure 12.4 Potentials $V(x) = |x|^n$ for $n = 2$, 6, 10, and 40. The infinite-square well of width 2 is shown for comparison. For n up to about 6, the breadth of the region where $V(x)$ is nearly zero increases substantially.

The following figures show another interesting case of symmetric one-dimensional problems. Here the potentials are a set: $V(x) = |x|^n$ for $n = 1, 2, 3, \ldots$. These potentials approximate the infinite square well but with rounded corners; the harmonic oscillator (with $\hbar = m = 1$ and $k = 2$) is a member of the set. Figure 12.4 shows (one-half of) several members of the set of potentials ($n = 2$, 6, 10, and 40). Figure 12.5 shows the ground-state and second-excited-state wave functions for $n = 2$, 10, and 40. Figure 12.6 shows the ground-state energy and the second-excited-state energy as functions of n. The ground-state eigenvalue has a minimum for $n \cong 6$. For $n \lesssim 6$, both the kinetic-energy and potential-energy terms in Schrödinger's equation drop as n increases because a larger and larger region has essentially $V(x) = 0$. After $n \approx 6$, the wave function gains little $V = 0$ region but gains potential energy quickly in the cutoff regions beyond $x = 1$.

All the calculations discussed so far were done with a program such as SCH [Programs 12.1(a), (b), (c)]. All symmetric one-dimensional Schrödinger equation problems can be solved with such a program. A higher-convergence version (using a fourth-order Runge-Kutta method for increased accuracy) is SCH4TH [Programs 12.2(a), (b)].

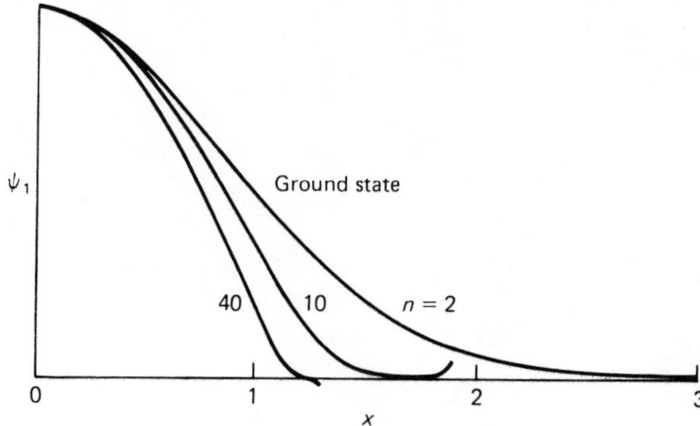

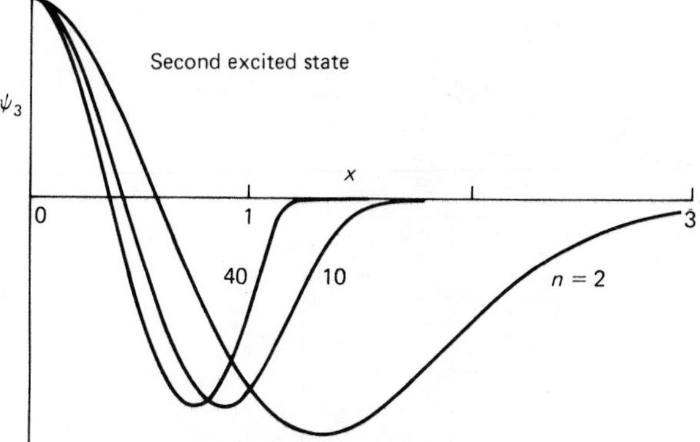

Figure 12.5 Several wave functions for the potentials $V(x) = |x|^n$. (Top) Ground-state wave functions for $n = 2$, 10, and 40. The traces are halted just as the wave functions start to diverge. (Bottom) Second-excited-state wave functions for $n = 2$, 10, and 40.

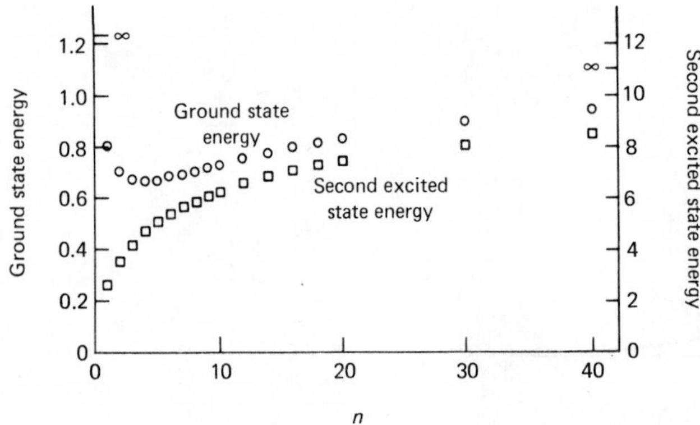

Figure 12.6 Energy levels as a function of the exponent n in the potential. The minimum in the ground-state energy near $n = 6$ is discussed in the text. Both curves are asymptotic to the infinite-square well ($n = \infty$).

Program 12.1 (a) Flow chart for a Schrödinger equation solution strategy. After setting initial parameters, you choose an energy. Schrödinger's equation then allows you to calculate the wave function a step at a time.

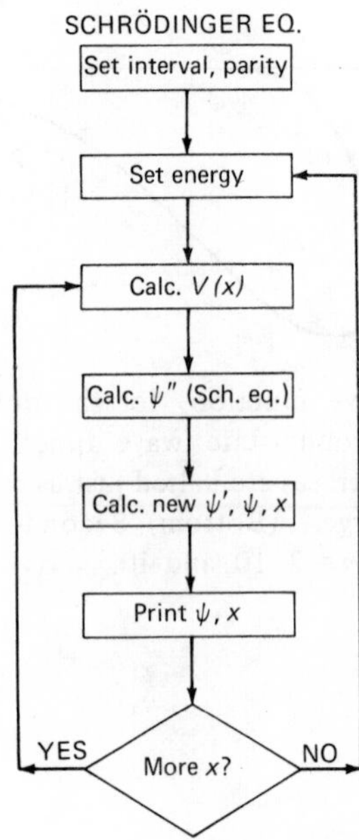

Program 12.1 (b)

```
SCHBAS    1D Schrödinger equation

100 DEF FNV(X)=X*X*X*X    Potential
110 PRINT "END OF INTERVAL, DELTA-X, PARITY (0=EVEN; 1=ODD), ENERGY?"
120 INPUT X9,X7,P,E
130 LET X0=0
140 LET X6=X9/20    Print distance
150 IF P=1 THEN 190
160 LET P0=1    Initial ψ    } Odd
170 LET P1=0    Initial ψ'   } parity
180 GOTO 210
190 LET P1=1    Initial ψ    } Even
200 LET P0=0    Initial ψ'   } parity
210 PRINT "X","P(X)","/P(X)/↑2"
220 LET P2=2*(FNV(X0+X7/2)-E)*(P0+P1*X7/2)    Schrödinger equation (half-stepped)
230 LET P0=P0+(P1+P2*X7/2)*X7    New ψ (half-stepped)
240 LET P1=P1+P2*X7    New ψ'
250 LET X0=X0+X7    New position
260 IF X0<X6 THEN 290    }
270 LET X6=X6+X9/20      } Print
280 PRINT X0,P0,P0*P0    } group
290 IF X0<X9 THEN 220    Return for next Δx step
300 PRINT
310 GOTO 110    Return for new guess
320 END
```

Program 12.1 (c)

```
SCHFOR    1D Schrödinger equation

      POTEN(X)=X*X*X*X    Potential
5     READ(5,101)ENDX,DX,PAR,E    End x, Δx, parity, energy
      IF (PAR-999.) 7,70,70
7     X=0.
      XINT=ENDX/20.    Print distance
      IF (PAR-1.) 10,20,10
10    PSI0=1.    Initial ψ    } Even
      PSI1=0.    Initial ψ'   } parity
      GOTO 30
20    PSI0=0.    Initial ψ    } Odd
      PSI1=1.    Initial ψ'   } parity
30    PSI2=2.*(POTEN(X+DX/2.)-E)*(PSI0+PSI1*DX/2.)    Schrödinger equation
                                                       (half-stepped)
      PSI0=PSI0+(PSI1+PSI2*DX/2.)*DX    New ψ
      PSI1=PSI1+PSI2*DX    New ψ'
      X=X+DX    New distance
      IF (X-XINT) 50,40,40            }
40    XINT=XINT+ENDX/20.              } Print
      PRINT 110,X,PSI0,PSI0*PSI0      } group
50    IF (X-ENDX) 30,60,60    Return for new Δx step
60    PRINT 120
      GOTO 5    Return for new guess
101   FORMAT(4F10.6)
110   FORMAT(1X,3E12.4)
120   FORMAT(1H )
70    STOP
      END
```

Program 12.2 (a)

SCH4THBA *1D Schrödinger equation (fourth-order Runge-Kutta)*

```
100 DEF FNV(X)=X*X*X*X    Potential
110 PRINT "END OF INTRVL, DELTA-X, PARITY (-1=ODD; +1=EVEN):"
120 INPUT L,D,P
130 IF P=1 THEN 170
140 LET P8=0    Initial ψ   } Odd
150 LET P9=1    Initial ψ'  } parity
160 GOTO 190
170 LET P8=1    Initial ψ   } Even
180 LET P9=0    Initial ψ'  } parity
190 PRINT "ENERGY";
200 INPUT E
210 PRINT "X","P(X)"
220 LET P0=P8
230 LET P1=P9
240 LET X=0
250 LET X1=0
260 PRINT X,P0
270 LET K1=P1
280 LET L1=2*(FNV(X)-E)*P0
290 LET K2=P1+L1*D/2
300 LET L2=2*(FNV(X+D/2)-E)*(P0+K1*D/2)   Fourth-order Runge-Kutta parameters
310 LET K3=P1+L2*D/2
320 LET L3=2*(FNV(X+D/2)-E)*(P0+K2*D/2)
330 LET K4=P1+L3*D
340 LET L4=2*(FNV(X+D)-E)*(P0+K3*D)
350 LET P0=P0+D*(K1+2*K2+2*K3+K4)/6    New ψ
360 LET P1=P1+D*(L1+2*L2+2*L3+L4)/6    New ψ'
370 LET X=X+D    New distance
380 LET X1=X1+D
390 IF X1<L/10 THEN 420   } Print
400 LET X1=0              } group
410 PRINT X,P0
420 IF X<L+D/10 THEN 270   Return for next Δx step
430 PRINT
440 GOTO 190    Return for new energy guess
450 END
```

12.4 Some Three-Dimensional Schrödinger Equation Problems

The full Schrödinger equation for stationary states is

$$-\frac{\hbar^2}{2m}\left(\frac{\partial^2\psi}{\partial x^2}+\frac{\partial^2\psi}{\partial y^2}+\frac{\partial^2\psi}{\partial z^2}\right)+V(\mathbf{r})\psi=E\psi$$

For many problems, the potential $V(\mathbf{r})$ is such that the equation is separable, that is, it can be separated into three total differential equations, each of which depends on only one variable. A simple example of this separation is the three-dimensional harmonic oscillator. This Schrödinger equation separates into three equations in x, y, and z. Each equation is identical to the harmonic oscillator discussed above. The full solution of the three-dimensional harmonic oscillator is the product of three one-dimen-

Program 12.2 (b)

```
SCH4THFO   1D Schrödinger equation (fourth-order Runge-Kutta)
           V(X)=X*X*X*X     Potential
           DATA ENDX,DX,PAR/3.,.01,1./
  10       READ(5,100)E     Energy
           IF (PAR.EQ.-1.) GOTO 20  ┐
           PSI=1.     ψ             │
           PSI1=0.    ψ'            │
           GOTO 30                  ├ Initial ψ, ψ'
  20       PSI=0.                   │
           PSI1=1.                  ┘
  30       X=0.
           PRINT 101,X,PSI
  40       PSI1A=PSI1                                           ┐
           PSI2A=2.*(V(X)-E)*PSI                                │
           PSI1B=PSI1+PSI2A*DX/2.                               │ Fourth-order
           PSI2B=2.*(V(X+DX/2.)-E)*(PSI+PSI1A*DX/2.)            ├ Runge-Kutta
           PSI1C=PSI1+PSI2B*DX/2.                               │ parameters
           PSI2C=2.*(V(X+DX/2.)-E)*(PSI+PSI1B*DX/2.)            │
           PSI1D=PSI1+PSI2C*DX                                  │
           PSI2D=2.*(V(X+DX)-E)*(PSI+PSI1C*DX)                  ┘
           PSI=PSI+DX*(PSI1A+2.*PSI1B+2.*PSI1C+PSI1D)/6.   New ψ
           PSI1=PSI1+DX*(PSI2A+2.*PSI2B+2.*PSI2C+PSI2D)/6.   New ψ'
           X=X+DX      New position
           PRINT 101,X,PSI
           IF (X-ENDX) 40,10,10
           STOP
  100      FORMAT(F8.3)
  101      FORMAT(1X,2E12.4)
           END
```

sional solutions—one in x, one in y, and one in z. When the potential $V(\mathbf{r})$ is spherically symmetric, that is, $V(\mathbf{r}) = V(|\mathbf{r}|)$, Schrödinger's equation separates in spherical coordinates (r, θ, ϕ). The total differential equations in θ and ϕ can be solved immediately, and the problem reduces to solving the radial equation

$$\frac{1}{r^2}\frac{d}{dr}\left(r^2\frac{dR}{dr}\right) = \left\{\frac{2m}{\hbar^2}\left[V(|\mathbf{r}|) - E\right] + \frac{l(l+1)}{r^2}\right\}R$$

where $l = 0, 1, 2, \ldots$ is a number specifying the angular momentum of the particle around the force center. This equation can be transformed to a simpler equation by substituting $P(r) = rR(r)$ and using Hartree's atomic units:

$$\frac{d^2P}{dr^2} = \left\{2\left[V(|\mathbf{r}|) - E\right] + \frac{l(l+1)}{r^2}\right\}P$$

The final radial equation can be solved very much as if it were a one-dimensional problem.

In Hartree atomic units, the unit of energy (e^2/a) is twice the ionization potential for hydrogen; the units of charge and mass are the magnitudes of the charge and mass

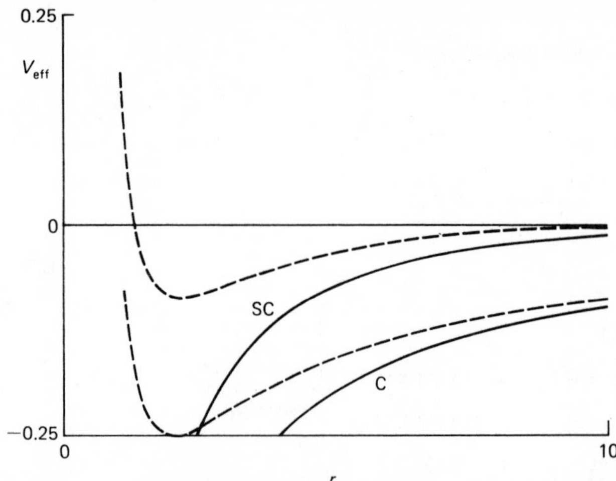

Figure 12.7 The effective potentials for the $l = 0$ and $l = 1$ states of the Coulomb (C) potential and the screened Coulomb (SC) potential. The screening parameter is a 5. A state cannot be bound if its energy lies wholly below the effective potential, $V(r) + l(l+1)/2r^2$. In the present units, the ground state for the hydrogen atom is $E_1 = -0.5$ and cannot be a bound state for the $l = 1$ or P state.

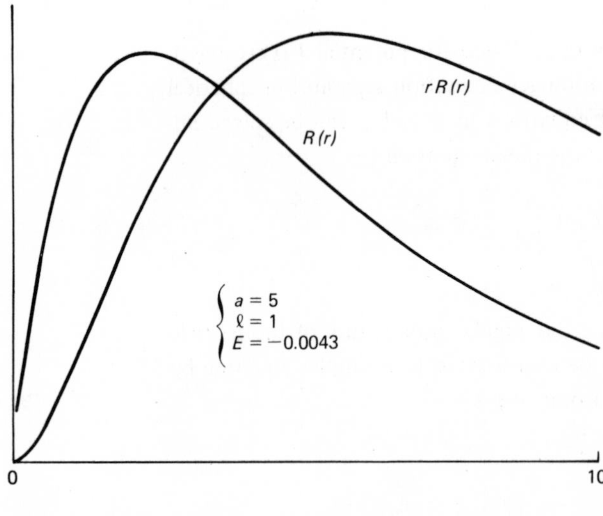

Figure 12.8 The radial wave functions $R(r)$ and $rR(r)$ and energy E for the lowest energy P state of the screened Coulomb potential. The energy is given in Hartree atomic units; the state is just barely bound. The potential is $\exp(-\frac{1}{5}r)/r$ in atomic units.

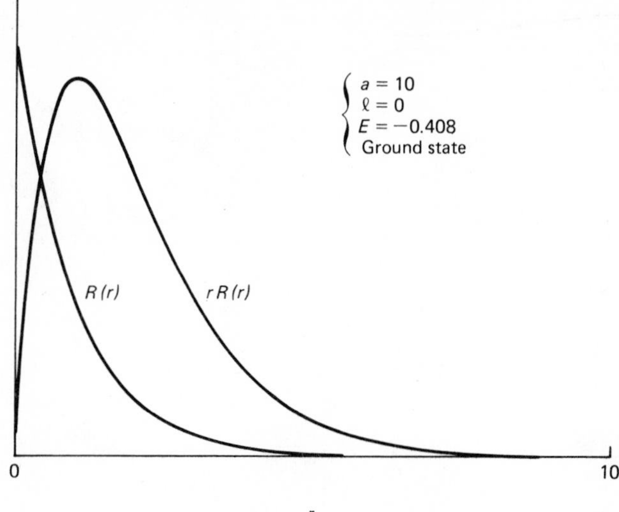

Figure 12.9 The radial wave functions and energy for the ground state of the screened Coulomb potential $-\exp(-\frac{1}{10}r)/r$. For comparison, the same state for the unscreened Coulomb potential has energy -0.5.

of the electron; the unit of length is the first Bohr radius ($a = \hbar^2/me^2$). (For a fuller account see H. A. Bethe and E. E. Salpeter, *Quantum Mechanics of One- and Two-Electron Atoms,* Academic Press, New York, 1957). The Coulomb potential energy becomes Z/r in the Schrödinger equation, and the energy levels for the hydrogen atom are $-1/(2n^2)$.

It is always useful to compare the numerical method with analytical solutions. If the numerical method is applied to the Coulomb potential, the energies of the first states agree to (better than) three-figure accuracy, and the wave functions appear indistinguishable from the closed-form solutions. The energies of the states do not depend on l, but the wave functions do. The $l = 1$ state (P state) has no energy below $E = -0.125$. The effective potential $[-1/r + l(l+1)/(2r^2)]$ for $l = 1$ has a minimum of -0.25. The ground state with $E = -0.5$ has $E < V_{\text{eff}}\ (l = 1)$ and a decaying solution everywhere when $l = 1$.

Another example is the screened Coulomb potential $V(\mathbf{r}) = -\exp(-r/a)/r$. Figure 12.7 compares the Coulomb and screened Coulomb effective potentials with $l = 0$ and 1 in the centripetal term $[l(l+1)/(2r^2)]$ and a screening parameter $a = 5$. Figure 12.8 shows the lowest energy state with $l = 1$ for the screened potential with $a = 5$. The energy is -0.0043. Figure 12.9 shows the ground state of a screened Coulomb potential with screening parameter $a = 10$. Figures 12.10 and 12.11 compare the $l = 0$ and

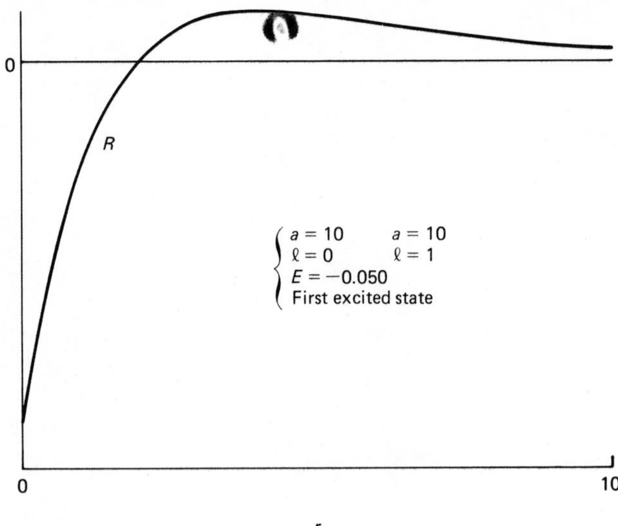

Figure 12.10 The radial wave functions and energy for the first excited S state in the screened Coulomb potential $-\exp(-\frac{1}{10}r)/r$. For comparison, the energy for the analogous state in the unscreened Coulomb potential has energy -0.125 and is degenerate with the lowest energy P state.

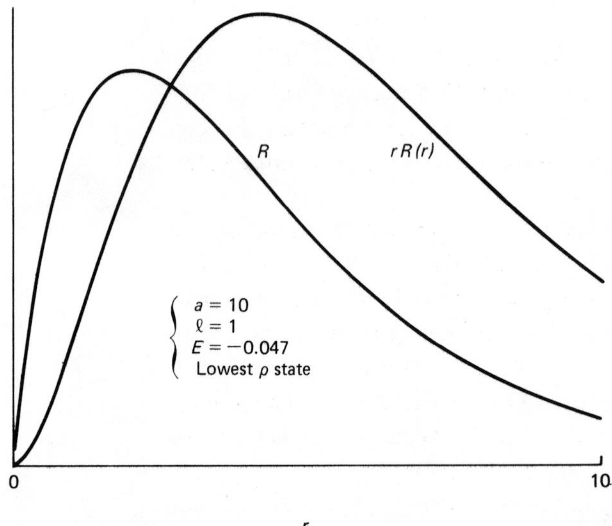

Figure 12.11 The radial wave function and energy for the lowest energy P state in the screened Coulomb potential $-\exp(-\frac{1}{10}r)/r$. The degeneracy between the $n = 2$, $l = 1$ and the $n = 2$, $l = 0$ states found in the hydrogen atom, is removed by the screening (see Figure 12.10).

Program 12.3 (a)

```
SCH3D1BA     3D Schrödinger equation (out from origin)

100 DEF FNV(R)=-EXP(-R/8)/R    Potential
110 PRINT "R(START),R(END),DELTA-R,ORBITAL #(L),ENERGY?"
120 INPUT R8,R9,R7,L,E
130 LET P0=R8↑(L+1)    Initial wave function, P = rR(r)
140 LET P1=(L+1)*R8↑L    Initial P' = dP/dr
150 LET R=R8
160 PRINT R,P0,P0/R
170 LET R5=(R9-R8)/20    Print distance
180 LET R6=R+R7/2
190 LET P2=2*(FNV(R6)+L*(L+1)/(2*R6*R6)-E)*(P0+P1*R7/2)    Schrödinger equation
200 LET P0=P0+(P1+P2*R7/2)*R7    New P (half-stepped)
210 LET P1=P1+P2*R7    New P'
220 LET R=R+R7    New distance
230 IF R<R5 THEN 260          \
240 LET R5=R5+(R9-R8)/20       } Print group
250 PRINT R,P0,P0/R           /
260 IF R<R9 THEN 180    Return for next Δr step
270 PRINT
280 GOTO 110    Return for new try
290 END
```

Program 12.3 (b)

```
SCH3D1FO     3D Schrödinger equation (out from origin)

          V(R)=-EXP(-R/8.)/R    Potential
10        READ(5,100)RST,REND,DR,AL,E    Starting r, ending r, delta r, l, energy
          L=AL
          IF (RST-999.) 20,90,20
20        P0=RST**(L+1)    Initial wave function, P = rR(r)
          P1=(AL+1.)*RST**L    Initial P' = dP/dr
          R=RST
          PRINT 102,AL,E
          PRINT 101,R,P0,P0/R
          PRNTR=(REND-RST)/20.    Print distance
25        RTEMP=R+DR/2.                                          Schrödinger
          P2=2.*(V(RTEMP)+AL*(AL+1.)/(2.*RTEMP**2)-E)*(P0+P1*DR/2.)  equation
          P0=P0+(P1+P2*DR/2.)*DR    New P (half-stepped)
          P1=P1+P2*DR    New P'
          R=R+DR    New distance
          IF (R-PRNTR) 40,30,30          \
30        PRNTR=PRNTR+(REND-RST)/20.      } Print group
          PRINT 101,R,P0,P0/R            /
40        IF (R-REND) 25,10,10    Return for next Δr step or new try
90        STOP
100       FORMAT(6F10.5)
101       FORMAT(1X,3E12.4)
102       FORMAT(1X,2HL=,F5.2,4H  E=,E12.4)
          END
```

$l = 1$ states of the screened Coulomb potential with $a = 10$ for $n = 2$. The energies are $E = -0.050$ and $E = -0.047$. The degeneracy in l of the energies for the Coulomb potential is broken with even a small amount of screening.

All the calculations were done with a program like SCH3D1 [Programs 12.3(a), (b)]. This program is very similar to SCH in that it initializes the wave function near the origin. Whenever the effective potential is dominated by the angular momentum term near $r = 0$, $P(r)$ near zero goes as r^{l+1}.

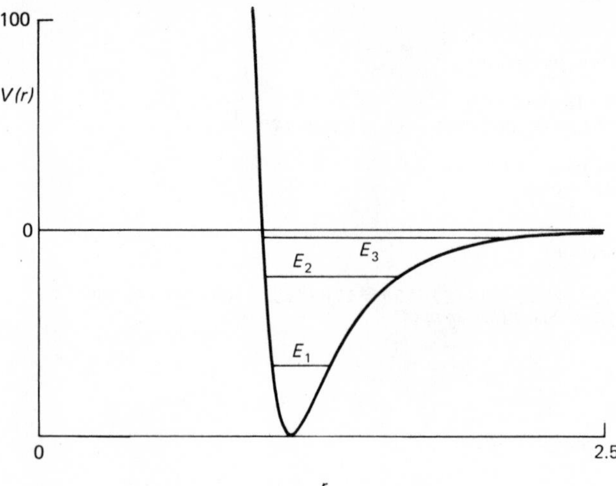

Figure 12.12 The Lennard-Jones 6–12 potential. The well depth is 100; the separation at the minimum is 1.1225; the potential is zero at $r = 1$. The first three bound states are indicated; a fourth state is just barely bound.

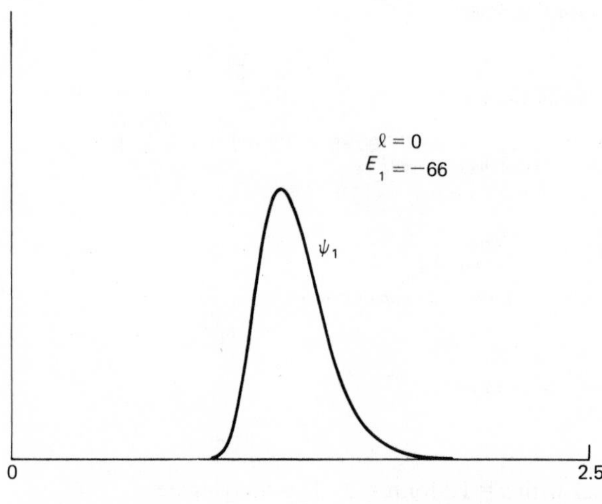

Figure 12.13 The ground-state wave function and energy for the 6–12 potential in Figure 12.12. Although the calculation is performed as a one-dimensional problem, the solution is the radial wave function in an S state for the three-dimensional problem.

12.5 Another Method for Three-Dimensional Potentials

When you cannot initialize the wave function near the origin, you can still use a simple numerical approach. Find a region (usually far from the origin) where you know an approximate solution for ψ and ψ' (asymptotic solutions). Start out there and integrate back in step by step. Use a boundary condition near $r = 0$ to find the eigenvalues.

One example is $V(r) = V_0(1/r^{12} - 1/r^6)$ (the 6–12 potential, important in molecular problems). For large r, $V(r) \approx 0$, and for energies less than zero (bound states), P is approximately proportional to $\exp(-\alpha x)$ with $\alpha = \sqrt{(2m/\hbar^2)(-E)}$. Then $P' = dP/dr = \alpha P$ and you can set P and P' to start the calculation correctly. Integrate step by step back toward $r = 0$ and remember that for an eigenvalue E, P must go to zero at $r = 0$ (since $V \longrightarrow \infty$). Again you watch the tail of P (now the tail near $r = 0$) whip through zero at eigenvalues.

Figure 12.12 is the 6–12 potential. Figure 12.13 is the ground-state wave function. Figure 12.14 is the first-excited-state wave function. Figure 12.15 is the second-excited-state wave function. All these calculations were performed by a program such as SCH3D2 [Programs 12.4(a), (b)].

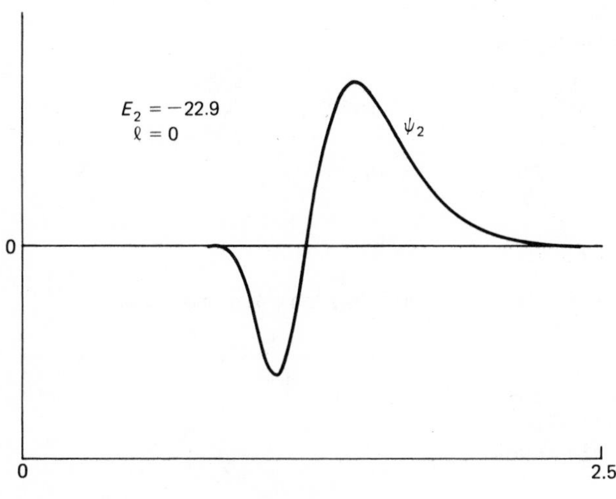

Figure 12.14 The first-excited-state wave function and energy for the 6–12 potential.

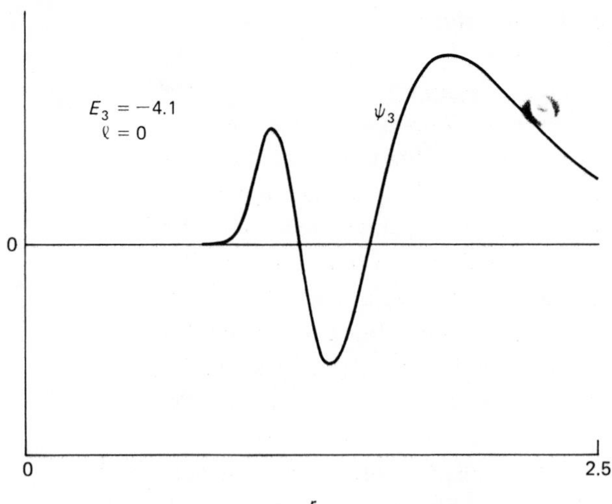

Figure 12.15 The second-excited-state wave function and energy for the 6–12 potential.

Program 12.4 (a)

```
SCH3D2BA     3D Schrödinger equation (in toward origin)

100 DEF FNV(R)=400*(R↑(-12)-R↑(-6))   Potential
110 PRINT "R(START),R(END),DELTA-R,ORBITAL #(L),ENERGY?"
120 INPUT R9,R8,R7,L,E
130 LET R7=-ABS(R7)
140 LET P0=EXP(-SQR(2*(FNV(R9)-E))*R9)   Initial P = rR(r)
150 LET P1=-SQR(2*(FNV(R9)-E))*P0   Initial P' = dP/dr
160 LET R=R9
170 PRINT R,P0,P0/R
180 LET R5=(R8-R9)/20 + R9   Print distance
190 LET R6=R+R7/2
200 LET P2=2*(FNV(R6)+L*(L+1)/(2*R6*R6)-E)*(P0+P1*R7/2)   Schrödinger equation
210 LET P0=P0+(P1+P2*R7/2)*R7   New P (half-stepped)
220 LET P1=P1+P2*R7   New P'
230 LET R=R+R7   New distance
240 IF R>R5 THEN 270          }
250 LET R5=R5+(R8-R9)/20      } Print group
260 PRINT R,P0,P0/R           }
270 IF R>R8 THEN 190   Return for next Δr step
280 PRINT
290 GOTO 110   Return for new try
300 END
```

Program 12.4 (b)

```
SCH3D2FO    3D Schrödinger equation (in toward origin)

            V(R)=400.*(1./R**12-1./R**6)  Potential
10          READ(5,100)RST,REND,DR,AL,E   Start r, end r, Δr, l, energy
            L=AL
            IF (RST-999.) 20,90,20
20          DR=-ABS(DR)
            P0=EXP(-SQRT(2.*(V(RST)-E))*RST)   Initial P = rR(r)
            P1=-SQRT(2.*(V(RST)-E))*P0   Initial P' = dP/dr
            R=RST
            PRINT 102,AL,E
            PRINT 101,R,P0,P0/R
            PRNTR=RST+(REND-RST)/20.    Print distance
25          RTEMP=R+DR/2.
            P2=2.*(V(RTEMP)+AL*(AL+1.)/(2.*RTEMP*RTEMP)-E)*(P0+P1*DR/2.)
            P0=P0+(P1+P2*DR/2.)*DR   New P (half-stepped)    Schrödinger equation
            P1=P1+P2*DR   New P'
            R=R+DR   New distance
            IF (R-PRNTR) 30,30,40         }
30          PRNTR=PRNTR+(REND-RST)/20.    } Print group
            PRINT 101,R,P0,P0/R           }
40          IF (R-REND) 10,10,25   Return for next Δr step or new try
90          STOP
100         FORMAT(6F10.5)
101         FORMAT(1X,3E12.4)
102         FORMAT(1X,2HL=,F5.2,4H  E=,E12.4)
            END
```

12.6 One-Dimensional Wave-Packet Propagation

The time-dependent Schrödinger equation in one dimension is

$$-\frac{\hbar^2}{2m}\frac{\partial^2 \Psi(x,t)}{\partial x^2} + V(x,t)\,\Psi(x,t) = i\hbar\,\frac{\partial\,\Psi(x,t)}{\partial t}$$

This equation can be solved directly using partial differential equation techniques, but when the potential is time-independent, a simpler method works. If the potential is independent of time $[V(x,t) = V(x)]$ then the equation separates by setting $\Psi(x,t) = \psi(x)\,T(t)$. The equation for $\psi(x)$ is the time-independent Schrödinger equation,

$$-\frac{\hbar^2}{2m}\frac{d^2\psi(x)}{dx^2} + V(x)\psi(x) = E\psi(x)$$

The equation for $T(t)$ is

$$\frac{dT(t)}{dt} = -\,i\frac{E}{\hbar}\,T(t)$$

E is the separation constant and plays the role of the energy.

The equation for $T(t)$ can be solved immediately to obtain $T(t) = Ae^{-iEt/\hbar} = Ae^{-iEt}$ in Hartree atomic units. The constant A is a normalization constant.

The spatial equation can be solved analytically (for a few problems) or numerically, as discussed previously. First, we will discuss the spatial part of the wave function for continuum states. Then we will put in the time part and propagate wave packets, pulses made up by summing many wave functions.

One-Dimensional Continuum States

Suppose we have some potential well such that for $|x| > x_0$, the potential $V(x)$ is essentially zero. Then, for $|x| > x_0$, the Schrödinger equation for the spatial wave function $\psi(x)$, becomes $\psi''(x) = -2E\,\psi(x)$. The solutions for $\psi(x)$ are $\psi(x) = \cos(kx)$, $\psi(x) = \sin(kx)$ for $E > 0$. Neither solution may be discarded because of boundary conditions; both solutions are well behaved everywhere. There are two solutions to the equation for every energy E. Furthermore, since there are no boundary conditions which limit the acceptable energies, every energy $E > 0$ is allowed. The pairs of states form a continuum.

If the potential $V(x)$ is symmetric, we can choose the states to be wholly even in x or wholly odd in x (if we wish). Since any linear combination of solutions to the Schrödinger equation is itself a solution, we could choose any arbitrary phase and make $\sin(kx + \theta)$ one of our two solutions; the other solution (by orthogonality) would then be $\cos(kx + \theta)$.

The simplest potential of the sort we have just been discussing is $V(x) = 0$ everywhere (the free particle). The even and odd wave functions for energy E in this potential are $\cos(kx)$ and $\sin(kx)$, where $k = \sqrt{2E}$ in atomic units. The full solutions $\psi(x, t)$ for the states of energy E are then $\psi(x, t) = \cos(kx)e^{-ik^2t/2}$ and $\sin(kx)e^{-ik^2t/2}$. We will now put together sets of these states to make pulses, or wave packets.

Free Wave Packets

In the continuum we have a large range of closely spaced allowed energies and wave vectors, so we can build up a shaped pulse. Such a pulse is usually known as a wave packet. To make a Gaussian-shaped wave packet, $\exp(-x^2/x_0^2)$, for example, you use a range of wave vectors k (around k = some central k_0) each with its own amplitude $A(k)$, where $A(k) \propto x_0 \exp[-\frac{1}{4}(k - k_0)^2 x_0^2]$. (This is a formula you can find in books on Fourier transforms.)

You can build up a free wave packet with almost any shape you wish. For a square wave packet with width x_0,

$$A(k) = x_0 \frac{\sin[\frac{1}{2}(k - k_0)x_0]}{\frac{1}{2}(k - k_0)x_0}$$

For a triangular wave packet (with width $2x_0$),

$$A(k) = x_0 \frac{\sin^2 [\frac{1}{2}(k - k_0) x_0]}{[\frac{1}{2}(k - k_0) x_0]^2}$$

The central wave vector k_0 determines the speed of the packet. The group velocity is $v = dE/dk = \hbar k_0/m$ or k_0 in Hartree units.

What happens when you set up such a wave packet at time zero and then let go? The packet shape will distort and deform as the higher-energy states in the packet shift out ahead of the lower-energy states.

The Propagation of Free Wave Packets

A program can easily calculate and add up the amplitudes of the various components of a packet. Because of limitations on how many amplitudes can be stored and how long it may take a program to execute, an approximation to the continuous distribution of amplitudes $A(k)$ must be made. Using 101 wave vectors across the Gaussian Fourier amplitude, for example, produces a good Gaussian wave packet in space.

We set up a packet, centered at the origin, at $t = 0$; we then let time pass and calculate each component's time factor, $\exp(-iEt)$. This part of the full wave function shifts the relative phases of the various components of the wave packet (since each component has its own energy E). At some later time you add the phase-shifted components back up. The packet has moved and distorted. Notice that the time parts of the wave functions do nothing to the amplitudes of the states because $|\exp(-iEt)| = 1$. The program PACKET [Programs 12.5(a), (b), (c)] implements this strategy.

Wave-Packet Propagation through a Square Well

Another case in which the spatial part of the wave function can be derived analytically is the finite-square well. D. S. Saxon, *Elementary Quantum Mechanics,* Holden-Day, San Francisco, 1968, gives the results (pages 147-152) as do many other quantum mechanics books. Film loops (which solve this problem in a more complicated way) are also available. The solution is often written in terms of (complex) amplitudes for right-going and left-going waves in the three regions: region I left of the well, region II in the well, and region III right of the well.

Region I	Region II	Region III
$\longrightarrow 1$	$\longrightarrow B1$	$\longrightarrow C$
$\longleftarrow B$	$\longleftarrow B2$	

The coefficients are found by equating the wave functions (and their first derivatives) at each boundary of the well.

Program 12.5 (a) Flow chart for a wave-packet-propagation strategy. After the initial shape of the packet and the time the packet has propagated is chosen, the strategy adds up the component wave functions at each point in space.

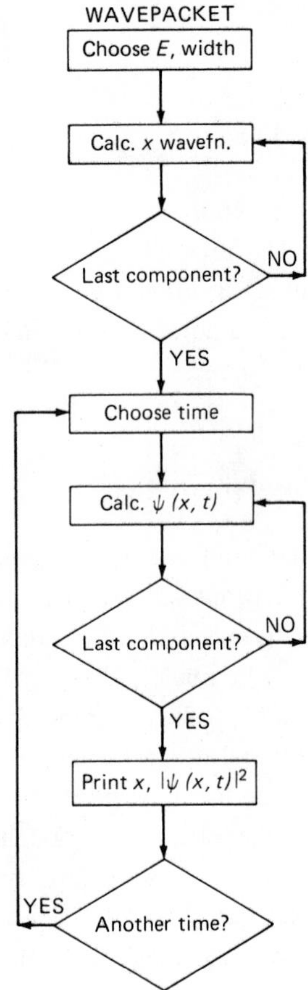

Program 12.5 (b)

```
PACKETBA      Free wave packet propagation

100 PRINT "CENTRAL ENERGY, PACKET WIDTH";
110 INPUT E0,X0
120 LET K0=SQR(2*E0)     Central wave vector
130 LET K9=5/(25*X0)     Wave vector increment
140 FOR I=1 TO 101                                              Store
150 LET K1=(I-51)*K9                                            components
160 LET K(I)=K0+K1     Wave vector                              of Gaussian
170 LET E(I)=K(I)*K(I)/2     Energy                             wave packet
180 LET A(I)=X0*1.25331*EXP(-K1*K1*X0*X0/8)   Gaussian amplitude
190 NEXT I
200 PRINT "LO K, HI K=",K(1),K(101)
210 PRINT "E(LO K), E(HI K)=",E(1),E(101)
220 PRINT "TIME";
230 INPUT T    Propagation time
240 IF T=999 THEN 400
250 FOR J=1 TO 25     Step across x
260 LET X=-2+(J-1)*.5
270 LET A1=.5*A(1)*COS(K(1)*X-E(1)*T)    Re(ψ)                  Sums for
280 LET A2=.5*A(1)*SIN(K(1)*X-E(1)*T)    Im(ψ)                  real and
290 FOR I=2 TO 100                                              imaginary
300 LET A1=A1+A(I)*COS(K(I)*X-E(I)*T)                           part of
310 LET A2=A2+A(I)*SIN(K(I)*X-E(I)*T)                           wave function
320 NEXT I                                                      of packet
330 LET A1=A1+.5*A(101)*COS(K(101)*X-E(101)*T)
340 LET A2=A2+.5*A(101)*SIN(K(101)*X-E(101)*T)
350 LET A1=A1*K9
360 LET A2=A2*K9
370 PRINT X,A1*A1+A2*A2     x, |ψ(x)|²
380 NEXT J
390 GOTO 220     Return for next propagation time
400 END
```

Program 12.5 (c)

```
PACKETFO     Free wave packet propagation

          DIMENSION AMPL(101),E(101)
          REAL K1,K(101),K0
          COMPLEX AMPLI
          DATA E0,DELTX/10.,1./     Central energy, packet width
          K0=SQRT(2.*E0)     Central wave vector
          DELTK=5./(25.*DELTX)     Wave vector increment
          DO 10 I=1,101                                          Store
          K1=FLOAT(I-51)*DELTK                                   components
          K(I)+K0+K1     Wave vector                             of Gaussian
          E(I)=K(I)*K(I)/2.     Energy                           wave
10        AMPL(I)=DELTX*1.25331*EXP(-K1*K1*DELTX*DELTX/8.)  Gaussian amplitude   packet
12        READ(5,101)T     Time packet propagated
          IF (T-999.) 14,40,40
14        DO 30 J=1,25     Step across x
          X=-2.+FLOAT(J-1)*.5                                    Sum for
          AMPLI=.5*AMPL(1)*CEXP(CMPLX(0.,K(1)*X-E(1)*T))         (complex)
          DO 20 I=2,100                                          wave
20        AMPLI=AMPLI+AMPL(I)*CEXP(CMPLX(0.,K(I)*X-E(I)*T))      function
          AMPLI=AMPLI+.5*AMPL(101)*CEXP(CMPLX(0.,K(101)*X-E(101)*T))   of packet, ψ
          AMPLI=AMPLI*DELTK
          AMPL1=REAL(AMPLI)
          AMPL2=AIMAG(AMPLI)
          AMPL3=AMPL1*AMPL1+AMPL2*AMPL2
30        PRINT 100,X,AMPL3     x, |ψ(x)|²
          GOTO 12     Return for next propagation time
40        STOP
100       FORMAT(1X,2F12.5)
101       FORMAT(F10.4)
          END
```

Program 12.6

PACKSQFO *Propagation of wave packet through finite square well*

```
          REAL K(51),KAP(51),K0
          COMPLEX A(51),PSIX,PSIXT,C,B,B1,B2,DENOM,ARG
          PI=3.14159265
          A0=.5
          V0=50.   Depth of well
          E0=50.   Central energy of packet
          K0=SQRT(2.*E0)   Central wave vector
          X0=-5.   Initial position of packet
          X1=1.5   Width of packet
          DELTK=5./(25.*X1)   Wave vector increment
          DO 10 I=1,51
          K(I)=FLOAT(I-26)*DELTK+K0    k    (For notation, see Saxon)
          KAP(I)=SQRT(2.*(K(I)*K(I)/2.+V0))   Kappa
          ARG=CMPLX(0.,-(K(I)-K0)*X0)
          AK=K(I)-K0
10        A(I)=X1*1.25331*EXP(-AK*AK*X1*X1/8.)*CEXP(ARG)   Gaussian centered at x0
20        READ(5,100)T   Time packet has propagated
          IF (T-999.) 15,50,15
15        DO 30 I=1,25   Step across x
          X=FLOAT(I-13)*.5
          PSIXT=CMPLX(0.,0.)
          DO 40 J=1,51   Sum over packet components
          AK=-(K(J)/KAP(J)+KAP(J)/K(J))
          DENOM=CMPLX(COS(2.*KAP(J)*A0),AK*SIN(2.*KAP(J)*A0)/2.)
          ARG=CMPLX(0.,-2.*K(J)*A0)
          C=CEXP(ARG)/DENOM
          IF (X+A0) 80,80,60
60        IF (X-A0) 70,90,90
70        ARG=CMPLX(0.,(K(J)-KAP(J))*A0)
          B1=(1.+K(J)/KAP(J))*CEXP(ARG)*C/2.
          ARG=CMPLX(0.,(K(J)+KAP(J))*A0)
          B2=(1.-K(J)/KAP(J))*CEXP(ARG)*C/2.     Coefficients of
          ARG=CMPLX(0.,KAP(J)*X)                 components when
          PSIX=B1*CEXP(ARG)                      -A0 < x < +A0
          ARG=-ARG
          PSIX=PSIX+B2*CEXP(ARG)
          GOTO 25
80        ARG=CMPLX(0.,-2.*K(J)*A0+PI/2.)
          B=.5*CEXP(ARG)*(KAP(J)/K(J)-K(J)/KAP(J))*SIN(2.*KAP(J)*A0)/DENOM
          ARG=CMPLX(0.,K(J)*X)
          PSIX=CEXP(ARG)          Coefficients of components when x <= -A0
          ARG=-ARG
          PSIX=PSIX+B*CEXP(ARG)
          GOTO 25
90        ARG=CMPLX(0.,K(J)*X)     Coefficients of
          PSIX=C*CEXP(ARG)         components when x >= +A0
25        E=K(J)*K(J)/2.   Energy of each component
          ARG=CMPLX(0.,-E*T)
40        PSIXT=PSIXT+A(J)*PSIX*CEXP(ARG)*DELTK   Sum for ψ(x, t) of packet
          PSI1=REAL(PSIXT)
          PSI2=AIMAG(PSIXT)
          PSI3=PSI1*PSI1+PSI2*PSI2
30        PRINT 101,X,PSI3   x, |ψ(x, t)|^2
100       FORMAT(F10.5)
101       FORMAT(1X,2F8.3)
          GOTO 20   Return for new propagation time
50        STOP
          END
```

[There is no BASIC program because most BASIC compilers do not allow complex numbers.]

Being able to calculate the spatial wave functions $\psi(x)$ for any x and for any energy E, you can follow the same strategy in forming and propagating wave packets. It is easiest if you form your wave packet completely to the left of the well (where the solutions are just the same sines and cosines discussed for free wave packets). One program implementing this strategy is PACKSQ [Program 12.6].

Propagation through any One-Dimensional Well

The spatial wave functions $\psi(x)$ for any energy E in the continuum can be computed algorithmically as discussed above. Hence, knowing the spatial wave functions for all the energies of the component wave functions in a wave packet, you can form and propagate packets past any such potential. The process is easiest if you center the initial wave packet around $x = 0$ and center the well off to the right. You can initialize the two wave functions to cosine and sine waves at the origin (which merely amounts to setting the overall phase). Form the packet at $x = 0$ to the left of the well; then start the packet toward the well. In the end, part of the packet will get past the well; part will be reflected. The smoother the potential, the easier it is for the packet to get past. Smooth edges on the potential also diminish the rapid oscillations (the interference effects) at the edges of the well.

One program to propagate wave packets past potential wells is VPAC [Program 12.7(a), (b)]. (The program uses a fourth-order Runge-Kutta method of integration to find the spatial wave function. Adding up wave functions to make packets demands that not only the amplitudes of the wave functions but also the phases must be very accurate.)

Because of the limitations of some computers, only 21 different wave vectors are used in the program presented. This compromise leads to "ghost" packets which image the main packet but lie at large positive and negative x. The compromise also means that the packet shape is only approximate. This last program takes a long time to execute. On some small machines it is a half hour before the program finishes finding all the spatial wave functions.

You can integrate Schrödinger's equation much as you did $\mathbf{F} = m\mathbf{a}$. The result is a knowledge of wave functions and (through $|\psi|^2$) the distributions of electrons in a number of interesting problems. The intuitive grasp of the microscopic behavior of particles and wave functions that you can gain will stand you in good stead through any further work in the physical sciences.

Problems

1. Find stationary-state solutions for finite square wells with $V_0 = 1000$ and $V_0 = 10$. Estimate how many states are bound in each case. Compare the

Program 12.7(a)

```
VPACBAS   Packet propagated past a potential, V(x)

100 DIM P(21,26),Q(21,26),E(21),C(21),S(21)
110 DEF FNV(X)=-100*EXP(-(X-6)*(X-6))   Potential
120 PRINT "CENTRAL ENERGY, HALF-WIDTH OF PACKET?"
130 INPUT E8,X6
140 LET K8=SQR(2*E8)   Wave vector for central energy
150 LET K9=3/(10*X6)   Wave vector increment
160 LET X1=.1   Step size, Δx
170 FOR I=1 TO 21   Store energies, wave functions for components
180 LET K0=K8+(I-11)*K9
190 LET E(I)=K0*K0/2   Energy
200 LET K7=K0-K8
210 LET A=.399569*X6*EXP(-X6*X6*K7*K7/2)
220 LET Q(I,1)=0
230 LET Q0=0
240 LET P1=0
250 LET Q1=K0*A
260 LET P(I,1)=A
270 LET P0=A
280 LET X0=0
290 LET X5=0
300 LET N=1
310 LET K1=P1
320 LET M1=Q1
330 LET L0=2*(FNV(X0)-E(I))
340 LET L1=L0*P0
350 LET N1=L0*Q0
360 LET K2=P1+L1*X1/2
370 LET M2=Q1+N1*X1/2
380 LET L0=2*(FNV(X0+X1/2)-E(I))
390 LET L2=L0*(P0+K1*X1/2)          Fourth-order
400 LET N2=L0*(Q0+M1*X1/2)          Runge-Kutta
410 LET K3=P1+L2*X1/2               parameters
420 LET M3=Q1+N2*X1/2
430 LET L3=L0*(P0+K2*X1/2)
440 LET N3=L0*(Q0+M2*X1/2)
450 LET K4=P1+L3*X1
460 LET M4=Q1+N3*X1
470 LET L0=2*(FNV(X0+X1)-E(I))
480 LET L4=L0*(P0+K3*X1)
490 LET N4=L0*(Q0+M3*X1)
500 LET P0=P0+X1*(K1+2*K2+2*K3+K4)/6   New Re(ψ)
510 LET P1=P1+X1*(L1+2*L2+2*L3+L4)/6   New Re(ψ')
520 LET Q0=Q0+X1*(M1+2*M2+2*M3+M4)/6   New Im(ψ)
530 LET Q1=Q1+X1*(N1+2*N2+2*N3+N4)/6   New Im(ψ')
540 LET X0=X0+X1 – New position
550 LET X5=X5+X1
560 IF ABS(X5-.5)>X1/2 THEN 610
570 LET X5=0                        Store
580 LET N=N+1                       group
590 LET Q(I,N)=Q0
600 LET P(I,N)=P0
610 IF X0<12.6 THEN 310   Return for next Δx step
620 NEXT I
630 PRINT "TIME THE PACKET HAS PROPAGATED?"
640 INPUT T0
650 FOR I=1 TO 21             Store
660 LET C(I)=COS(E(I)*T0)     sines and
670 LET S(I)=SIN(-E(I)*T0)    cosines
680 NEXT I
690 FOR J=1 TO 26   Step across x
700 LET X0=(J-1)*.5
710 LET P1=0
720 LET P2=0
730 FOR I=1 TO 21   Sum up components
740 LET P1=P1+K9*(P(I,J)*C(I)-Q(I,J)*S(I))   Re(ψ)
750 LET P2=P2+K9*(Q(I,J)*C(I)+P(I,J)*S(I))   Im(ψ)
750 NEXT I
770 PRINT X0,P1*P1+P2*P2
780 NEXT J
790 PRINT
800 GOTO 630   Return for new propagation time
810 END
```

Program 12.7(b)

```
VPACFOR      Packet propagated past a potential, V(x)

             DIMENSION E(21)
             COMPLEX C(21),PSI,P(21,26),P0,P1,K1,K2,K3,K4,L1,L2,L3,L4
             V(X)=-100.*EXP(-(X-6.)*(X-6.))   Potential
             DATA EN,WID/50.,1./   Central energy, width of packet
             CENK=SQRT(2.*EN)   Wave vector for central energy
             DK=3./(10.*WID)   Wave vector increment
             DX=.1   Step size, Δx
             DO 30 I=1,21   Store energies, wave functions of components
             AK=(I-11)*DK
             E(I)=(AK+CENK)**2/2.   Energy
             A=.399569*WID*EXP(-WID*WID*AK*AK/2.)
             P(I,1)=CMPLX(A,0.)
             P0=P(I,1)
             P1=CMPLX(0.,(AK+CENK)*A)
             X0=0.
             PRNTX=0.
             N=1
5            K1=P1                                     ⎫
             L1=2.*(V(X0)-E(I))*P0                     ⎪
             K2=P1+L1*DX/2.                            ⎪
             AL0=2.*(V(X0+DX/2.)-E(I))                 ⎪ Fourth-order Runge-
             L2=AL0*(P0+K1*DX/2.)                      ⎬ Kutta parameters
             K3=P1+L2*DX/2.                            ⎪
             L3=AL0*(P0+K2*DX/2.)                      ⎪
             K4=P1+L3*DX                               ⎪
             L4=2.*(V(X0+DX)-E(I))*(P0+K3*DX)          ⎭
             P0=P0+DX*(K1+2.*K2+2.*K3+K4)/6.    New component ψ
             P1=P1+DX*(L1+2.*L2+2.*L3+L4)/6.    New component ψ'
             X0=X0+DX   New x
             PRNTX=PRNTX+DX                            ⎫
             IF (ABS(PRNTX-.5)-DX/2.) 15,15,20         ⎪ Store
15           PRNTX=0.                                  ⎬ group
             N=N+1                                     ⎪
             P(I,N)=P0                                 ⎭
20           IF (X0-12.6) 5,30,30   Return for next Δx step
30           CONTINUE
40           READ(5,100)T0   Time pulse has propagated
             IF (T0-999.) 45,70,70
45           DO 50 I=1,21   Store sines and cosines
50           C(I)=CMPLX(COS(E(I)*T0),SIN(-E(I)*T0))
             DO 60 J=1,26   Step across x
             X0=(J-1)*.5
             PSI=(0.,0.)
             DO 55 I=1,21   Sum over components
55           PSI=PSI+DK*P(I,J)*C(I)
             PSI2=REAL(PSI)**2+AIMAG(PSI)**2
60           PRINT 101,X0,PSI2
             GOTO 40   Return for new time
70           STOP
100          FORMAT(F10.5)
101          FORMAT(1X,F8.3,F10.5)
             END
```

energies of the first couple states to the energies of the corresponding states in the infinite square well.

2. Find stationary-state solutions for potential $V(x) = x^4$. There are no analytic solutions. Find the first three wave functions and energies. Compare the energies to those of the harmonic oscillator. Explain any differences physically.

3. Find some stationary states for a finite well with rounded corners: $V(x) = 50x^6$, $|x| < 1$; $V(x) = 100 - 50/x^6$; $|x| > 1$. Plot the potential V and find the ground-state eigenstate. Estimate the highest energy state bound.

13

nuclear decay and the ecology of isolated regions

13.1 Introduction

This chapter is concerned with one application of the computer to introductory nuclear physics–radioactive decay. It is easy to treat both single-decay systems and multiple-decay nuclear chains. The analytic solution is easy for a single-decay system but not for anything more complicated. Usually an analytic solution to a radioactive decay chain is discussed only if the half-lives of the products are very different, in which case the members dominate the whole system at widely different times. The numerical solution is easy for any set of half-lives. Even multiple-daughter decays with nearly equal half-lives are straightforward to handle.

The fundamental equation is a simple rate formula. For a single decay to a stable daughter, the rate at which the number N_1 of nuclei changes with time is proportional to the number of nuclei itself:

$$\frac{dN_1}{dt} = CN_1$$

where $C < 0$ for nuclear decay.

The next more-complicated system allows the daughter to decay with its own half-life. The coupled differential rate equations for this system are

$$\frac{dN_1}{dt} = C_1 N_1$$

$$\frac{dN_2}{dt} = C_2 N_2 + C_3 N_1$$

where $C_1 < 0$, $C_2 < 0$, and $C_3 = -C_1$ for nuclear decay. The analytic solution to this problem is very complicated; the numerical solution is easy, especially with a computer, even for the general case of N decaying nuclei in the chain.

An analogous problem occurs in the ecology of isolated regions. Here you want to calculate the numbers N_i of various species in the region as they reproduce and interact. The strategy of the numerical solution is very similar to that for nuclear-decay chains. Similar mathematical tools occur in many different areas of science. A short discussion of the isolated-region ecology problem is given at the end of the chapter.

13.2 Solutions of Nuclear-Decay Problems

The stable-daughter problem is the simplest decay problem to handle. The rate equation is $dN/dt = CN$ and can be solved both analytically and numerically. This affords a test of the numerical method used.

The rate equation can be solved as shown in the program NUCDEC [Program 13.1(a), (b), (c)]. The system being treated is the decay of a thorium isotope ($^{232}_{90}$Th) to radium and finally to lead. $^{232}_{90}$Th decays to $^{228}_{88}$Ra by emitting an α

Program 13.1(a) Flow chart for a nuclear-decay strategy. After setting the initial parameters, the strategy calculates the change in the number of decaying nuclei (and the new number) in each time step t.

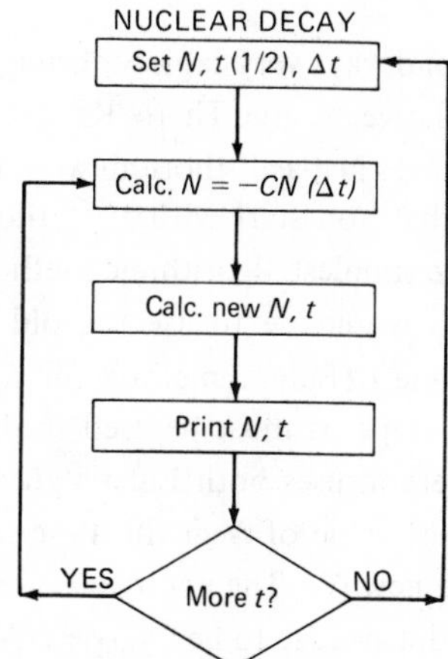

Program 13.1(b)

```
NUCDECBA     Simple nuclear decay

100 LET N0=1E22     Initial number of nuclei
110 LET N=N0
120 LET C=-.693/140     Decay constant
130 LET D=1     Time step, Δt
140 LET T0=10     Final time
150 PRINT "T(100KYR)","N(TH232)","EXPONENTIAL"
160 LET N=N+C*(N+C*N*D/2)*D     New number of nuclei (half-stepped)
170 LET T=T+D     New time
180 IF T<T0 THEN 210          }
190 PRINT T,N,N0*EXP(C*T)     } Print group
200 LET T0=T0+10              }
210 IF T<=200 THEN 160     Return for next Δt step
220 END
```

Program 13.1(c)

```
NUCDECFO     Simple nuclear decay

          AN0=1.E22     Initial number of nuclei
          AN=AN0
          C=-.693/140.     Decay constant
          DELT=1.     Time step, Δt
          PRNTT=10.     Print time
5         AN=AN+C*(AN+C*AN*DELT/2.)*DELT     New number of nuclei (half-stepped)
          T=T+DELT     New time
          IF (T-PRNTT) 20,10,10     }
10        ANTHE=AN0*EXP(C*T)        }
          PRINT 100,AN,ANTHE        } Print group
          PRNTT=PRNTT+10.           }
20        IF (T-200.) 5,5,30     Return for next Δt step
30        STOP
100       FORMAT(1X,2E12.6)
          END
```

particle ($^4_2$He). The radium decays very rapidly through a whole chain to $^{208}_{82}$Pb. These last decays are so fast relative to the Th to Ra decay (whose half-life is 1.4×10^{10} years) that the system behaves just as if thorium went directly to lead.

The program assumes that you start with 10^{22} thorium nuclei at zero time (about 1 g of pure thorium). The simplest algorithmic method (stepping along through time) to solve the rate equation would be to use the old value of the number N to find $\Delta N = CN_{\text{old}}\, \Delta t$ at each time t (Euler's method for first-order differential equations). However, very small time steps Δt must be used to derive good answers by means of this method. A better method uses both Euler's guess at the new N and the old N to find (a guess at) the average value of N in this time step. The average value is used to find the final value of the new N. The approach is related to trapezoidal rule integration. The program takes the average to be $(N_{\text{old}} + \Delta N_{\text{Euler}})/2 = N + CN\, \Delta t/2$ and uses this value to update N. The program also prints out values from the analytic solution.

Program 13.2(a)

```
DAUGHTBA     Decay of mother-daughter nuclear decay chain

100 LET N1=1E22   Initial number of nuclei
110 LET C1=-.693/2.2E3   }
120 LET C2=-.693/1.62E2  } Decay constants
130 LET C3=-C1           }
140 LET D=1     Time step, Δt
150 LET T0=10       Final time
160 PRINT "T(KYR)","N(NP237)","N(U233)"
170 LET L1=N1+C1*N1*D              } Estimate of new
180 LET L2=N2+C2*N2*D+C3*N1*D      } numbers of nuclei
190 LET N2=N2+C2*(N2+L2)*D/2+C3*(N1+L1)*D/2  } New numbers of nuclei
200 LET N1=N1+C1*(N1+L1)*D/2                 }
210 LET T=T+D    New time
220 IF T<T0 THEN 250   } Print
230 PRINT T,N1,N2      } group
240 LET T0=T0+10       }
250 IF T<200 THEN 170    Return for next Δt step
260 END
```

Program 13.2(b)

```
DAUGHTFO     Decay of mother-daughter nuclear decay chain

          AN1=1.E22   } Initial numbers of nuclei
          AN2=0.      }
          C12=-.693/2.2E3    }
          C23=-.693/1.62E2   } Decay constants
          C21=-C12           }
          DELT=1.   Time step, Δt
          PRNTT=10.   Final time
10        ANES1=AN1+C12*AN1*DELT                     } Estimate of new
          ANES2=AN2+C23*AN2*DELT+C21*AN1*DELT        } numbers of nuclei
          AN2=AN2+C23*(AN2+ANES2)*DELT/2.+C21*(AN1+ANES1)*DELT/2.  } New numbers
          AN1=AN1+C12*(AN1+ANES1)*DELT/2.                          } of nuclei
          T=T+DELT    New time
          IF (T-PRNTT) 20,15,15   } Print
15        PRINT 100,T,AN1,AN2     } group
          PRNTT=PRNTT+10.         }
20        IF (T-100.) 10,10,25   Return for next Δt step
25        STOP
100       FORMAT(1X,3E10.3)
          END
```

For a simple single decay you can integrate the rate equation directly and get

$$N = N_0 e^{Ct}$$

where N_0 is the initial number of nuclei. An execution of the program will show you that the numerical procedure is quite accurate. On most machines, the numerical solution is correct to five figures over several half-lives.

You can also carry these numerical ideas to the situation in which the daughter nucleus decays. First you use Euler's method to get a first approximation to the new numbers; then you use the average numbers to find good approximations to the next

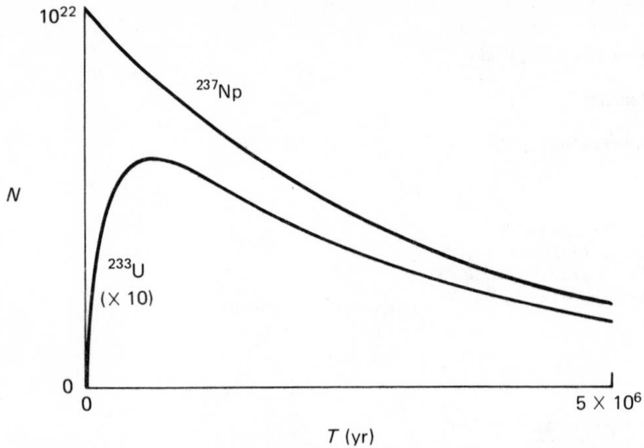

Figure 13.1 A single decaying-daughter nuclear decay chain. The decay of ^{237}Np through ^{233}U. The sample is initially pure with 10^{22} Np nuclei. The Np half-life is 2.2×10^6 years; the U half-life is 1.62×10^5 years. The intermediate nucleus ^{233}Pa decays to ^{233}U with a half-life of 27.4 days. The U curve is multiplied by 10.

values of the N's. The program DAUGHT [Programs 13.2(a), (b)] uses the following sequence of steps. The L's are Euler's approximations.

$$L_1 = N_1 + C_1 N_1 D$$

$$L_2 = N_2 + C_2 N_2 D + C_3 N_1 D$$

$$N_2 = N_2 + C_2 \left(\frac{N_2 + L_2}{2}\right) D + C_3 \left(\frac{N_1 + L_1}{2}\right) D$$

$$N_1 = N_1 + C_1 \left(\frac{N_1 + L_1}{2}\right) D$$

$$T = T + D$$

D is the time step Δt. The program DAUGHT illustrates this method for the nuclear decay of a neptunium isotope $^{237}_{93}$Np which decays to $^{233}_{91}$Pa which in turn decays to $^{229}_{90}$Th via $^{233}_{92}$U. All the other decays in the chain (ending at $^{209}_{83}$Bi which is stable) are very fast. The Np half-life is 1.62×10^5 years. The program assumes an initially pure sample of 10^{22} nuclei.

Figure 13.1 shows the numbers of neptunium and uranium nuclei as functions of time. Since the two dominant half-lives are a factor of 10 apart, analytical approximate solutions can be derived. The numbers are always dominated by the longer half-life.

Program 13.3(a)

```
U234BAS     Uranium-234 decay

100 LET N1=1E22     Initial number of nuclei
110 LET C1=-.693/250  ┐
120 LET C2=-.693/80   │
130 LET B2=-C1        ├ Decay constants
140 LET C3=-.693/1.62 │
150 LET B3=-C2        ┘
160 LET D=.01     Time step, Δt
170 LET T0=1     Print time
180 PRINT "T(KYR)","N(U234)","N(TH230)","N(RA226)"
190 LET L1=N1+C1*N1*D                  ┐ First estimates of
200 LET L2=N2+C2*N2*D+B2*N1*D          ├ new numbers of
210 LET L3=N3+C3*N3*D+B3*N2*D          ┘ nuclei
220 LET N3=N3+C3*(L3+N3)*D/2+B3*(L2+N2)*D/2  ┐ New
230 LET N2=N2+C2*(L2+N2)*D/2+B2*(L1+N1)*D/2  ├ numbers
240 LET N1=N1+C1*(L1+N1)*D/2                 ┘ of nuclei
250 LET T=T+D   New time
260 IF T<T0 THEN 290  ┐ Print
270 PRINT T,N1,N2,N3  ├ group
280 LET T0=T0+1       ┘
290 IF T<=20 THEN 190    Return for next Δt step
300 END
```

Program 13.3(b)

```
U234FOR     Uranium-234 decay

          AN1=1.E22     Initial number of 234 U nuclei
          C1=-.693/250.  ┐
          C2=-.693/80.   │
          B2=-C1         ├ Decay
          C3=-.693/1.62  │ constants
          B3=-C2         ┘
          DELT=.01     Time step, Δt
          PRNTT=1.     Print time
10        AN1ES=AN1+C1*AN1*DELT                 ┐ First estimate
          AN2ES=AN2+C2*AN2*DELT+B2*AN1*DELT     ├ of new numbers
          AN3ES=AN3+C3*AN3*DELT+B3*AN2*DELT     ┘ of nuclei
          AN3=AN3+C3*(AN3ES+AN3)*DELT/2.+B3*(AN2ES+AN2)*DELT/2.  ┐ New numbers
          AN2=AN2+C2*(AN2ES+AN2)*DELT/2.+B2*(AN1ES+AN1)*DELT/2.  ├ of nuclei
          AN1=AN1+C1*(AN1ES+AN1)*DELT/2.                          ┘
          T=T+DELT     New time
          IF (T-PRNTT) 20,15,15        ┐ Print
15        PRINT 100,T,AN1,AN2,AN3      ├ group
          PRNTT=PRNTT+1.               ┘
20        IF (T-20.) 10,30,30     Return for next Δt step
30        STOP
100       FORMAT(3X,F5.2,3(2X,E10.4))
          END
```

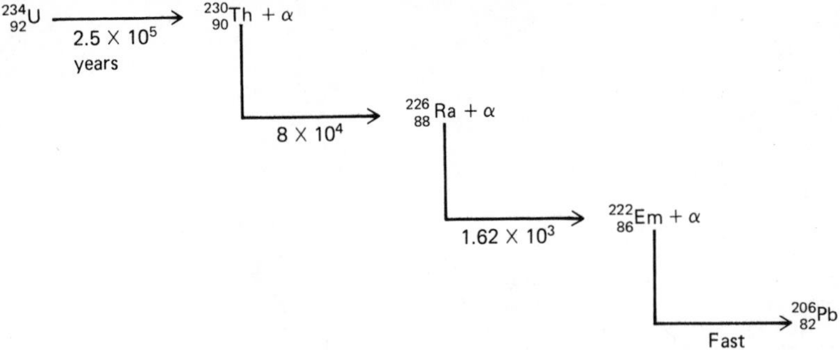

Figure 13.2 The nuclear decay scheme of $^{234}_{92}$U.

The numerical method is not limited to two species. The program U234 [Programs 13.3(a), (b)] illustrates the (dominant) decays in the uranium 234 chain. This is (essentially) a three-nuclei chain. The program assumes that you start with about a gram of pure ^{234}U. The decay scheme is shown in Figure 13.2. Again you find an Euler approximation to the new numbers (and then an average) before calculating the new values of the N's.

Figure 13.3 shows the number of U nuclei, the number of Th nuclei, and 100 times the number of Ra nuclei as functions of time. The number of Em nuclei, which is essentially the number of Pb nuclei, would be the initial number of nuclei minus the sum of the numbers of the three decaying constituents or 10^{22}: $(N_1 + N_2 + N_3)$. In nature, ^{234}U is produced from ^{238}U (the most abundant form of uranium) by a decay with a (dominant) half-life of 4.51×10^9 years; this half-life is so long that the ^{234}U chain can be separately considered assuming the only way the number of ^{234}U nuclei changes is by decay to thorium.

The system of rate equations for N species which interact by decay as a chain can be written as a matrix equation. The numerical solution can then be written in terms of matrices, as in the program MATDEC [Program 13.4]. The number of different kinds of nuclei is limited only by the storage of the computer. The particular system treated in MATDEC is the ^{234}U chain. The answers are those of the earlier program; the MATDEC program takes longer on some computers because it uses the BASIC MAT package. On the other hand, the program is not limited to a specific number of species in the chain.

One use of these programs is to illustrate the way nuclear chains work; another use is to reduce experimental data. If the decays in a chain have widely separated half-lives, you can usually find a span of time in which each decay dominates. Then you can find the half-life and subtract the correct number of decay counts of that process at each time from the original data. You then move to the next time region and find

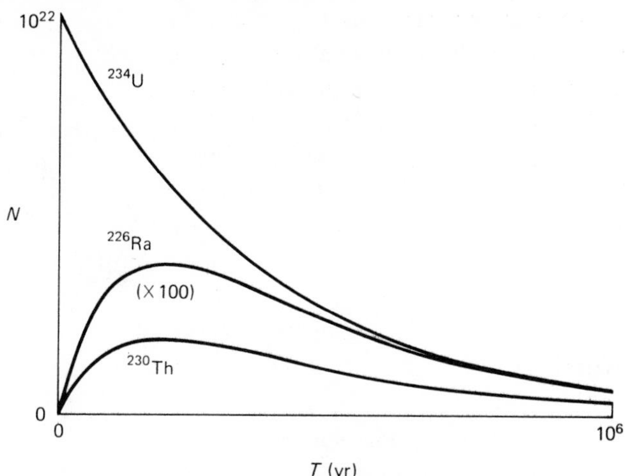

Figure 13.3 A two decaying-daughter nuclear chain. The decay of ^{234}U through ^{230}Th and ^{226}Ra. The half-lives are U, 2.5×10^5 years; Th, 8×10^4 years; Ra, 1.62×10^3 years. The Ra curve is multiplied by 100.

Program 13.4

```
MATDECBA    Nuclear decay chain

100 MAT READ N(3)          } Initial numbers
110 DATA 1E22,0,0          } of nuclei
120 MAT READ C(3,3)                                                  } Decay
130 DATA -.002772,0,0,+.002772,-8.66E-3,0,0,+8.66E-3,-.4278          } coefficients
140 LET D=.01        Time step, Δt
150 LET T0=1         Print time
160 MAT C=(D)*C
170 MAT M=C*N              } Estimate of
180 MAT M=(.5)*M           } half-stepped
190 MAT M=N+M              } numbers of nuclei
200 MAT L=C*M        Change in numbers of nuclei
210 MAT N=N+L        New numbers of nuclei
220 LET T=T+D        New time
230 IF T<T0 THEN 280       }
240 PRINT T;               }
250 MAT PRINT N,           } Print
260 PRINT                  } group
270 LET T0=T0+1            }
280 IF T<=20 THEN 170      Return for next Δt step
290 END
```

[There is no FORTRAN program because of differences between matrix manipulations on different computers.]

the next half-life, etc. This method works for widely separated half-lives. Reducing experimental data is accomplished by the numerical method when half-lives are not well separated. The process becomes one of trial and error. Using a guess at the right half-lives and initial numbers, you find how well the calculations fit the data. Then you correct your numerical values. It helps to have decent first guesses to start the process; sometimes the traditional dominant half-life method discussed above is used as a start. One advantage of the trial-and-error method is that you gain a feeling for how each parameter changes the results.

13.3 Nuclear Decay Using Random Numbers

Nuclear decay affords an interesting example of the use of random numbers to simulate statistical events. Such uses of random numbers are often called Monte Carlo calculations. A nuclear-decay problem using a Monte Carlo technique is given in the program NUMON [Programs 13.5(a), (b)]. Briefly, the calculation considers *N* nuclei and examines each nucleus after each unit of time. By comparing a random number to the decay constant, the program decides whether or not each nucleus decayed in the last unit of time. If the random number is less than the decay constant, the program decays the nucleus; if the random number (between 0 and 1) is equal to or larger than the decay constant, the nucleus is left alone. The random number simulates a random statistical process, so that examining how many nuclei decay as time progresses simulates the nuclear-decay process itself.

Program 13.5(a)

```
NUMONBA     Monte Carlo nuclear decay

100 DIM N(1000)
110 LET L(1)=.2          } Decay
120 LET L(2)=.05         } constants, λ
130 LET L(3)=0           }
140 FOR N=1 TO 1000      } Initial states
150 LET N(N)=1           } of nuclei
160 NEXT N               }
170 FOR T=1 TO 20      Step through time
180 FOR N=1 TO 1000    Step through nuclei
190 IF RND>=L(N(N)) THEN 210    Compare random number to λ
200 LET N(N)=N(N)+1     Decay nucleus
210 NEXT N
220 LET M(1)=0
230 LET M(2)=0
240 LET M(3)=0
250 FOR N=1 TO 1000    Step through nuclei
260 LET M(N(N))=M(N(N))+1     Count number in each type
270 NEXT N
280 PRINT T,M(1),M(2),M(3)
290 NEXT T
300 END
```

Program 13.5(b)

NUMONFO *Monte Carlo nuclear decay*

```
      DIMENSION NUC(1000),NUM(3),ALAM(3)
      ALAM(1)=.2      }Decay
      ALAM(2)=.05     }constants
      ALAM(3)=0.
      DO 20 I=1,1000    }Initialize states
20    NUC(I)=1          }of nuclei
      DO 90 J=1,20   Step through time
      DO 40 N=1,1000     Step through nuclei
      NN=NUC(N)
      IF (RANF(0)-ALAM(NN)) 30,40,40   Compare random number to decay constant
30    NUC(N)=NUC(N)+1     Flip spin
40    CONTINUE
      NUM(1)=0
      NUM(2)=0
      NUM(3)=0
      DO 80 N=1,1000    Step through nuclei
      NN=NUC(N)
80    NUM(NN)=NUM(NN)+1      Count number in each type
90    PRINT 101,J,NUM(1),NUM(2),NUM(3)
      STOP
101   FORMAT(1X,4I7)
      END
```

13.4 Ecology of Isolated Regions

The ecological equations for regions which do not communicate with the rest of the world are quite similar to the rate equations for nuclear-decay problems. The basic mathematical equations for two interacting species are the Lotka-Volterra equations.

$$\frac{dN_1}{dt} = R_1 N_1 \frac{K_1 - N_1 - J_1 N_2}{K_1}$$

$$\frac{dN_2}{dt} = R_2 N_2 \frac{K_2 - N_2 - J_2 N_1}{K_2}$$

The R_i's are the "biotic potentials," which are the unrestricted growth-rate constants. The K_i's are the "carrying capacities," which are the maximum number of each species that the region can support. The J_i's are the coefficients of species interaction, which vary from positive for inhibitory influences (such as one species eating the other) to negative for effects which increase numbers.

Rather than do a complete derivation, let us consider what reasonable terms in an ecological rate equation might look like. First, consider a species in an infinite environment (able to support an arbitrarily large number of individuals) with no species competition at all. The rate at which the numbers in the species will increase is then just proportional to the number of mature adults. For simplicity, assume that

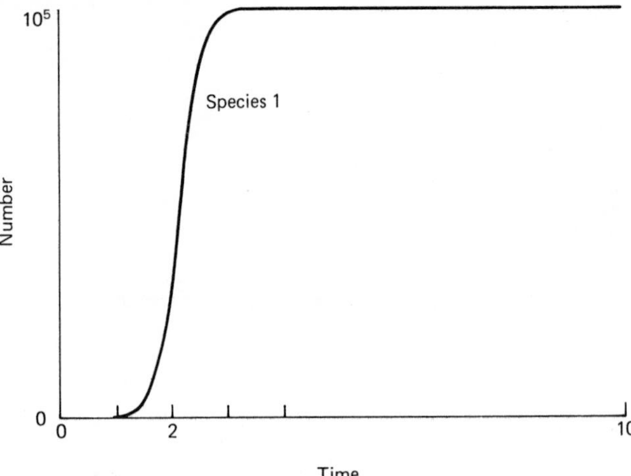

Figure 13.4 Growth of a single ecological species. A biotic potential (uninhibited growth rate) of 5, a carrying capacity (maximum number) of 10^5, and an initial number of 2 are used. No interactions with other species are included. By 4 years the maximum number has been reached.

maturation is instantaneous, so that the number of adults is the total number of individuals. Then,

$$\frac{dN_1}{dt} = C_1 N_1 \qquad \text{where } C_1 > 0 \text{ now}$$

As a next approximation, you know that any area has a limited maximum number of individuals it can support (the carrying capacity for the region). Call this maximum number M_1. You want to reflect this maximum in your equation. A very simple way to force an upper limit would be to make the rate of change of the number go to zero as the number approaches the maximum. This usual approximation is simply

$$\frac{dN_1}{dt} = C_1 N_1 \left(1 - \frac{N_1}{M_1}\right)$$

For $N_1 << M_1$ the growth is unaffected. As N_1 approaches M_1, the rate goes to zero, and the curve of N_1 versus t saturates. Figure 13.4 shows a species with its maximum number $M_1 = 10^5$ and a growth rate of five new individuals per old individual per unit time. (This might be a zeroth approximation to rabbits.)

This approximation to the way carrying capacity influences the growth is very simple, but it is neither very well validated by experiment nor very easy to believe.

The approach to saturation is too smooth. Actually you would expect the number to overshoot the maximum, turn over because there is not enough food (or something), drop back below the maximum, and then return in an oscillatory sort of a way. Presumably the oscillations slowly die out and a true saturation occurs. (Qualitatively, you expect the system to be somewhat "underdamped" rather than always "overdamped.") With these comments in mind, you can consider the simple form of the saturating curve-rate equation as the beginning of a series expansion. We have guessed the terms proportional to N_1 and N_1^2; other terms may be important in many systems. However, any model is hard to test. It is very hard to find a truly isolated region and often quite hard to get a good count of numbers. There is still qualitative agreement with the idea of a carrying capacity.

A third approximation would include competition between species. There are a large number of different forms the interaction can take and therefore, in principle, a large number of equations for different kinds of interaction. The simplest forms of interaction are usually proportional to the numbers of individuals in both species (unlike decay problems). An interaction should depend on the species "meeting," and the probability of meeting should depend on how many of each species there are. Write the interaction term as $A_1N_1N_2$, and then the system of equations for two competitive species is

$$\frac{dN_1}{dt} = C_1N_1\left(1 - \frac{N_1}{M_1}\right) + A_1N_1N_2 \qquad \frac{dN_2}{dt} = C_2N_2\left(1 - \frac{N_2}{M_2}\right) + A_2N_2N_1$$

C_i and M_i are positive; A_i can have either sign. This form of the equation is merely that of Lotka-Volterra if you write $C_i = R_i, M_i = K_i$, and $A_i = -J_iR_i/K_i$. In the ecology case (unlike the nuclear-decay case), there is no necessary relationship between A_1 and A_2. It may not help the fox as much to eat a rabbit as it hurts the rabbit to be eaten. Or, at least, it may not aid the number of foxes as much as it hurts the number of rabbits. This is particularly true if the fox has lots of other things he can eat but prefers a rabbit when he can find one.

The program ECOL [Programs 13.6(a), (b)] implements the strategy as developed for the nuclear-decay problem but applied to the ecology situation. The program as written assumes that you start with two members of each species and have growth rates of five and one per year. The maximum numbers are 10^5 and 10^2, respectively, and the interaction of species 2 on species 1 is -0.1. There is no interaction of species 1 on species 2. The program also assumes that once a species is depleted ($N < 1$), it does not recover.

Figure 13.5 shows the results for numbers of species 1 and 2 as functions of time. At first the growth rate dominates, and species 1 grows. Finally, however, the negative interaction takes over, and species 1 is annihilated.

Other effects can easily be included in programs such as ECOL. The age to maturation can be included by having the growth term depend on the number of individuals

Program 13.6(a)

ECOLBAS *Predator-prey calculations*

```
100 READ N1,N2                                    Initial numbers in two species
110 DATA 2,2
120 READ R1,M1,R2,M2,A1,A2                        Rate coefficients
130 DATA 5,1E5,1,100,-.1,0
140 LET D=.01                                     Time step, Δt
150 LET T0=.499                                   Print time
160 LET L1=N1+(R1*N1*(1-N1/M1)+A1*N1*N2)*D        First estimate of new
170 LET L2=N2+(R2*N2*(1-N2/M2)+A2*N2*N1)*D        numbers in species
180 LET L1=(L1+N1)/2                              Estimates of average
190 LET L2=(L2+N2)/2                              numbers in this Δt
200 LET N1=N1+(R1*L1*(1-L1/M1)+A1*L1*L2)*D        Second estimate of new
210 IF N1>1 THEN 230                              numbers in species
220 LET N1=0                                      (210-220: Species dead when only one left)
230 LET N2=N2+(R2*L2*(1-L2/M2)+A2*L2*L1)*D
240 IF N2>1 THEN 260
250 LET N2=0
260 LET T=T+D                                     New time
270 IF T<=T0 THEN 300                             Print
280 PRINT T,N1,N2                                 group
290 LET T0=T0+.5
300 IF T<=10 THEN 160                             Return for next Δt step
310 END
```

ECOLFOR *Predator-prey calculations*

```
      DATA AN1,AN2/2.,2./          Initial numbers in two species
      DATA GRO1,AMX1,GRO2,AMX2,EAT1,EAT2/5.,1.E5,1.,100.,-.1,0./     Rate coefficients
      DELT=.1                      Time step, Δt
      T=0.
      PRNTT=.499                   Print time
5     AN1ES=AN1+(GRO1*AN1*(1-AN1/AMX1)+EAT1*AN1*AN2)*DELT      First estimate of
      AN2ES=AN2+(GRO2*AN2*(1-AN2/AMX2)+EAT2*AN2*AN1)*DELT      new number in species
      AN1ES=(AN1ES+AN1)/2.         Estimate of average
      AN2ES=(AN2ES+AN2)/2.         numbers in this Δt
      AN1=AN1+(GRO1*AN1ES*(1-AN1ES/AMX1)+EAT1*AN1ES*AN2ES)*DELT    Second
      IF (AN1-1.) 10,10,20         Species dead when               estimate
10    AN1=0.                       only one left                   of new
20    AN2=AN2+(GRO2*AN2ES*(1-AN2ES/AMX2)+EAT2*AN2ES*AN1ES)*DELT    numbers
      IF (AN2-1.) 30,30,40                                         in species
30    AN2=0.
40    T=T+DELT                     New time
      IF (T-PRNTT) 60,60,50        Print
50    PRINT 100,T,AN1,AN2          group
      PRNTT=PRNTT+.5
60    IF (T-10.) 5,5,70            Return for next Δt step
70    STOP
100   FORMAT(1X,F10.4,2(2X,F10.4))
      END
```

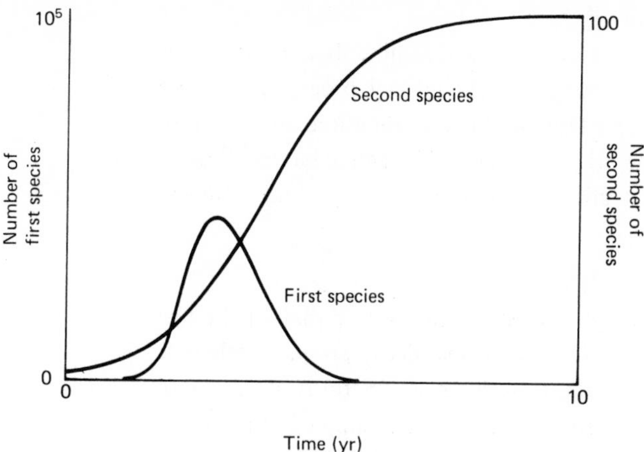

Figure 13.5 Growths of two interacting ecological species. Species 1 is the same as that used for Figure 13.3 but with an inhibitory interaction (coefficient = −0.1) with species 2. Species 2 has an initial number of 2, a maximum number of 100, and an unhibited growth rate of 1. By 7 years, species 2 has consumed all of species 1.

above a certain age. Random or cyclical environmental effects can be taken into account; more-complicated forms of interaction can be treated; multiple interactions can be treated; and better forms of approach to saturation can be discovered.

There are simple ways to handle rate equations using the computer. It is straightforward to handle situations such as chains of nuclear decay or isolated-region ecology. Numerical integration makes the problems easy even when analytical solutions are impossible.

Problems

The following are examples of ways to use the procedures introduced in this chapter.

1. a. Write down the decay chain for $^{238}_{92}$U.
   b. Pick out the longest decay time in the chain and assume that all the rest of the decays are instantaneous. Using a nuclear-decay program, calculate the numbers of the various important products as functions of time until you have only one-tenth of the original (pure) $^{238}_{92}$U sample left as $^{238}_{92}$U.
   c. Now include both the longest and next-to-longest decay times. Repeat part b.

2. a. Write down the decay chain for $^{235}_{92}U$.
   b. Pick out the longest decay time in the chain and assume that all the rest of the decays are instantaneous. Using a nuclear-decay program, calculate the numbers of the various important products as functions of time until you have only one-tenth of the original (pure) $^{235}_{92}U$ sample left as $^{235}_{92}U$.
   c. Now include both the longest and next-to-longest decay times. Repeat part b.

3. a. Write down the decay chain for $^{237}_{93}Np$.
   b. Pick out the longest decay time in the chain and assume that all the rest of the decays are instantaneous. Using a nuclear-decay program, calculate the numbers of the various important products as functions of time until you have only one-tenth of the original (pure) $^{237}_{93}Np$ sample left as $^{237}_{93}Np$.
   c. Now include both the longest and next-to-longest decay times. Repeat part b.

14 solid-state physics

14.1 Introduction

There are a number of interesting examples of uses of the computer in introductory solid-state physics (a course customarily taken at the senior undergraduate level). The computer expands your range of experience with various important aspects of introductory solid-state physics. This chapter discusses four uses: a simple way to draw Brillouin-zone constructions and Harrison free-electron Fermi surfaces; a way to calculate the propagation of phonon pulses down a one-dimensional chain; a way to display ω-versus-q diagrams (dispersion relations) and calculate densities of states for two- (or more-) dimensional lattices; and a way to calculate energy-versus-k surfaces for nearly free electrons in two (or more) dimensions.

14.2 The Reciprocal Lattice

It is useful to draw one Brillouin-zone construction in the reciprocal lattice by hand. The Brillouin construction for a simple two-dimensional square lattice is a good choice. Once you have seen how the construction is performed, the computer can relieve you of a lot of the arithmetic. It is helpful to see many examples; a few examples of Brillouin zone and Harrison constructions are given in Figures 14.1 to 14.4.

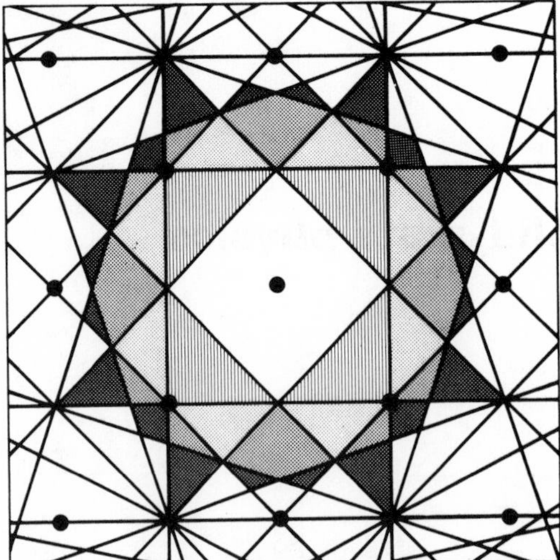

Figure 14.1 Brillouin-zone construction for a centered square two-dimensional lattice. Enough bisectors are drawn to complete pieces through the seventh zone; the second, third, fourth, and fifth zones are shaded.

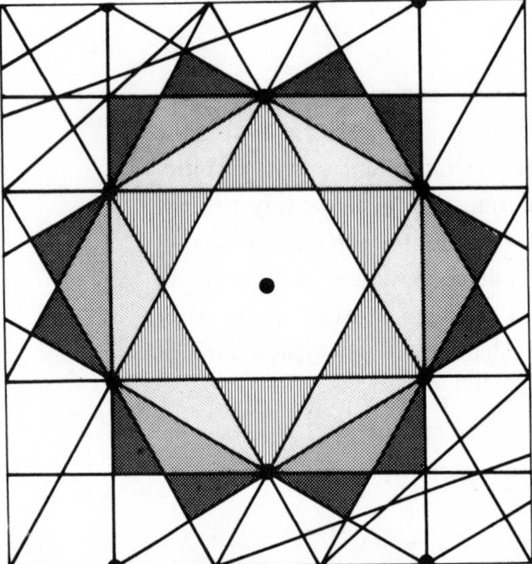

Figure 14.2 Brillouin-zone construction for a hexagonal two-dimensional lattice. Enough bisectors are drawn to complete the fifth zone; the second, third, fourth, and fifth zones are shaded.

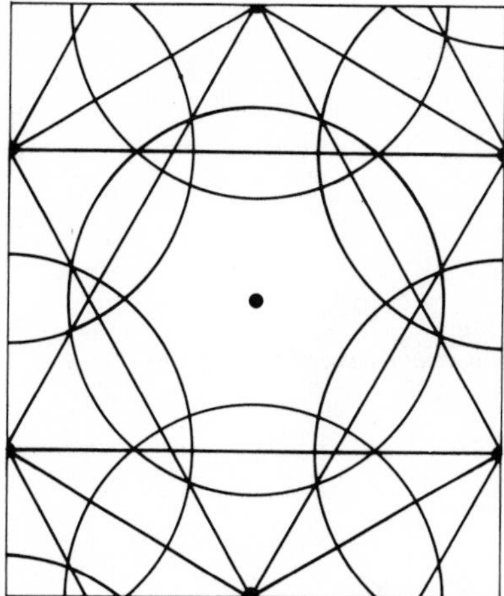

Figure 14.3 Brillouin-zone construction for a centered rectangular lattice with Harrison-construction circles for three electrons per cell. The Fermi surface has a hole piece in the second zone and small electron pieces in the third zone.

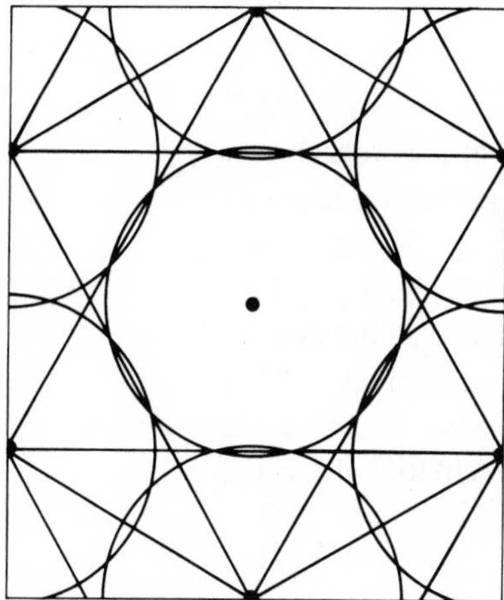

Figure 14.4 Brillouin-zone and Harrison-construction for a centered rectangular lattice with two electrons per cell. The Fermi surface has a small hole piece in the first zone and small electron pieces in the second zone.

Program 14.1(a)

HARRBAS *Reciprocal lattice and Harrison construction (2D)*

```
100 PRINT "2 DIRECT LATTICE VECTORS (X1,Y1,X2,Y2)."
110 INPUT X2,Y2,X3,Y3
120 LET A=ABS(X2*Y3-Y2*X3)       Area of direct lattice cell
130 LET X0=2*3.14159*Y3/A        }
140 LET Y0=2*3.14159*(-X3)/A     } Reciprocal lattice
150 LET X1=2*3.14159*(-Y2)/A     } vectors
160 LET Y1=2*3.14159*X2/A        }
170 LET B=4*3.14159*3.14159/A    Area of reciprocal lattice cell
180 LET I0=INT(ABS(5*3.14159/(X0-X1)))  Number of important reciprocal lattice points
190 FOR I=-I0 TO I0              } Step through reciprocal lattice points
200 FOR J=-I0 TO I0              }
210 LET X5=I*X0+J*X1             } Midpoint between origin
220 LET Y5=I*Y0+J*Y1             } and reciprocal lattice point
230 PRINT "RL POINT:",X5,Y5
240 PRINT "BISECTOR:"
250 IF Y5<>0 THEN 280            }
260 PRINT X5,"-5",X5,"+5"        } Vertical
270 GOTO 320                     } bisector
280 LET S=-X5/Y5                 }
290 LET Y8=Y5/2+S*(-5-X5/2)      } Nonvertical
300 LET Y9=Y5/2+S*(5-X5/2)       } bisector
310 PRINT "-5",Y8,"+5",Y9        }
320 NEXT J
330 NEXT I
340 PRINT "# OF ELECTRONS/UNIT CELL (OR 0 TO STOP)";
350 INPUT N0
360 IF N0=0 THEN 410
370 LET N0=N0/A                  Number of electrons per unit area
380 LET R=SQR(2*3.14159*N0)      Radius of Harrison circle
390 PRINT "FERMI RADIUS =";R
400 GOTO 340
410 END
```

The program to calculate the necessary perpendicular bisectors of reciprocal lattice vectors is easy to write. Given a set of (primitive) direct lattice basis vectors (x_1, y_1), you calculate the (primitive) reciprocal lattice basis vectors (x_2, y_2). The reciprocal lattice points are then simply integral multiples of the basis vectors (ix_2, jy_2). If the slope of the line connecting each reciprocal lattice point to the origin is S, the slope of the perpendicular bisector (in two dimensions) is the negative reciprocal $-1/S$. Any two points on the line going through $(\frac{1}{2}ix_2, \frac{1}{2}jy_2)$ with slope $-1/S$ define that Brillouin boundary. One program which does these calculations is HARR [Programs 14.1(a), (b)].

14.3 One-Dimensional Phonon Propagation

It is sometimes hard to grasp the meaning of the dispersive properties of a one-dimensional linear chain (the simplest lattice-vibration calculation). What does it really mean to have the group velocity v_s = zero at the Brillouin-zone boundary when the phase velocity $v_p = \omega/k$ does not? Even when the medium is not absorptive,

Program 14.1(b)

```
HARRFOR    Reciprocal lattice and Harrison construction (2D)

      DATA X1D,Y1D,X2D,Y2D/1.,0.,0.,1./    Direct lattice vectors
      A=ABS(X1D*Y2D-Y1D*X2D)  Area of direct lattice cell
      X1R=2.*3.14159*Y2D/A
      Y1R=2.*3.14159*(-X2D)/A     Reciprocal lattice
      X2R=2.*3.14159*(-Y1D)/A     vectors
      Y2R=2.*3.14159*X1D/A
      B=4.*3.14159*3.14159/A   Area of reciprocal lattice cell
      I0=2*INT(ABS(5.*3.14159/(X1R-X2R)))-1    Number of important reciprocal
      DO 10 I=1,I0                             lattice points
      AI=I-I0          Step through reciprocal lattice points
      DO 10 J=1,I0
      AJ=J-I0
      XRLP=AI*X1R+AJ*X2R    Midpoint between origin and
      YRLP=AI*Y1R+AJ*Y2R    reciprocal lattice point
      PRINT 100,XRLP,YRLP
      IF (YRLP) 20,15,20    Vertical
15    PRINT 104,XRLP,XRLP   bisector
      GOTO 10
20    S=-XRLP/YRLP    Slope
      YLFT=YRLP/2.+S*(-5.-XRLP/2.)   Nonvertical
      YRT=YRLP/2.+S*(5.-XRLP/2.)     bisector
      PRINT 101,YLFT,YRT
10    CONTINUE
25    READ(5,100)AN   Number of electrons per unit cell
      IF (AN) 40,90,40
40    AN=AN/A    Number of electrons per unit area
      R=SQRT(2.*3.14159*AN)    Radius of Harrison circle
      PRINT 103,AN,R
      GOTO 25
90    STOP
100   FORMAT(1X,2F10.4)
101   FORMAT(1X,4H-5   ,F10.4,7H    +5   ,F10.4)
102   FORMAT(F10.5)
103   FORMAT(1X,2HN=,F8.2,16H   FERMI RADIUS= ,F10.4)
104   FORMAT(1X,F10.4,7H    -5   ,F10.4,7H    +5   )
      END
```

it is hard to see what really happens. Using a computer you can calculate what happens to a pulse of lattice waves (a phonon) as it propagates.

Set up a pulse as, say, a Gaussian pulse initially by looking up the Gaussian Fourier transform. Then decide how long (in time) the pulse is to have propagated. Move each component (each individual lattice wave in the pulse) the right amount, $v_p t$, where v_p is the correct phase velocity for that particular lattice wave. Finally, add all the moved lattice waves. The result is the pulse after propagating for time t.

If all the component lattice waves making up the pulse have the same phase velocity v_p, then they all move together, and the pulse moves down the chain unchanged. This happens when the pulse has components only from the continuum region near the center of the Brillouin zone. If, however, some of the lattice-wave components (those near the zone edge) have a smaller phase velocity, the slower modes will tend to fall behind. When the components are added up at the later time, the pulse shape is distorted and the group velocity (roughly, the speed at which the center of the pulse moves) will be slower.

Program 14.2(a)

```
PHONBAS    Propagation of phonon pulse down 1D chain

100 DIM F(97),V(97)
110 PRINT "LATTICE SPACING (A), CENTRAL K, & WIDTH IN K:"
120 INPUT A,K9,K8
130 LET P=3.14159    π
140 LET I0=0
150 FOR K0=K9-3*K8 TO K9+3*K8 STEP K8/16
160 LET I0=I0+1
170 LET F(I0)=EXP(-(K0-K9)*(K0-K9)/(K8*K8))    Gaussian pulse
180 IF ABS(K0)>P/A THEN 230
190 IF K0<>0 THEN 210
200 LET K0=1E-6
210 LET V(I0)=ABS(SIN(K0*A/2))/K0    v(k) in first zone
220 GOTO 250
230 LET K=K0-INT((K0+P/A)/(2*P/A))*(2*P/A)    } v(k) reduced
240 LET V(I0)=ABS(SIN(K*A/2))/K               } back to first zone
250 NEXT K0
    (lines 150-250: Store amplitudes and velocities of packet components)
260 PRINT "TIME OF PROPAGATION";
270 INPUT T1
280 FOR X1=-10 TO 10 STEP 20/64    Step across x
290 LET I1=0
300 LET I2=0
310 LET I0=0
320 FOR K0=K9-3*K8 TO K9+3*K8 STEP K8/16    Sum over components
330 LET I0=I0+1
340 LET C0=COS(K0*(X1-V(I0)*T1))
350 LET I1=I1+C0*F(I0)    Real part of pulse
360 LET S0=SIN(K0*(X1-V(I0)*T1))
370 LET I2=I2-S0*F(I0)    Imaginary part
380 NEXT K0
390 LET I1=I1/32
400 LET I2=I2/32
405 PRINT X1,I1,I2    x, Re(pulse at x), Im(pulse at x)
410 NEXT X1
420 GOTO 260    Return for next propagation time
430 END
```

If some components have wave vectors beyond the Brillouin-zone edge, they actually have velocities in the opposite direction (by periodicity, the velocity of the mode one reciprocal lattice vector away). When the central wave vector of the pulse is at the zone edge, the pulse distorts as time progresses, but the center of the pulse tends to stay put. Hence, v_g becomes zero.

Figure 10.12 shows these effects. A program implementing this phonon-propagation strategy is PHON [Programs 14.2(a), (b)].

14.4 Two-Dimensional Lattice Vibrations

Another example of the use of computers in introductory solid-state physics is the calculation of lattice waves in two dimensions, which is also an illustration of how analytic solutions and computer calculations can be helpful together. The computer displays results derived analytically and aids one's visualization of the solution.

Program 14.2(b)

```
PHONFOR     Propagation of phonon pulse down 1D chain

       REAL K,K8,K9,I1,I2,K0
       DIMENSION F(97),V(97)
       DATA A,K9,K8/3.14159,.5,1./   Lattice spacing, central k, width in k
       PI=3.14159
       DO 10 I0=1,97
       K0=K9-3.*K8+(I0-1)*K8/16.
       F(I0)=EXP(-(K0-K9)*(K0-K9)/(K8*K8))   Gaussian
       IF (ABS(K0)-PI/A) 3,3,7
3      IF (K0) 5,4,5
4      K0=1.E-6
5      V(I0)=ABS(SIN(K0*A/2.))/K0   v(k) in first zone
       GOTO 10
7      K=K0-INT((K0+PI/A)/(2.*PI/A))*(2.*PI/A)   }v(k) reduced back
       V(I0)=ABS(SIN(K*A/2.))/K                  }to first zone
10     CONTINUE
20     READ(5,100)T1   Time pulse propagated
       IF (T1-999.) 25,50,50
25     DO 40 IX=1,11   Step across x
       X1=-10.+(IX-1)*2.
       I1=0.
       I2=0.
       DO 30 I0=1,97   Sum over components
       K0=K9-3.*K8+(I0-1)*K8/16.
       C0=COS(K0*(X1-V(I0)*T1))
       I1=I1+C0*F(I0)   Real part
       S0=SIN(K0*(X1-V(I0)*T1))
30     I2=I2-S0*F(I0)   Imaginary part
       I1=I1/32.
       I2=I2/32.
40     PRINT 101,X1,I1,I2   x, Re(pulse at x), Im(pulse at x)
       GOTO 20   Return for next propagation time
50     STOP
100    FORMAT(F10.5)
101    FORMAT(1X,3E12.4)
       END
```

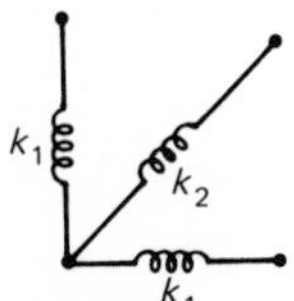

Figure 14.5 The geometry of the nearest- and next-nearest-neighbor forces in a two-dimensional square lattice.

Consider a two-dimensional square array of (equal) masses with nearest and next-nearest neighbors connected by springs in the x, y, and xy directions (see Figure 14.5). For convenience, let the springs in the x and the y directions be the same (k_1); but let the springs in the xy direction (next-nearest-neighbor springs) have a different spring constant (k_2). You can write down the equations of motion in x and y for any mass

Program 14.3(a)

```
OMOFQBAS        ω(q) for 2D square lattice

100 DIM W(11,11),D(26)
110 PRINT "LATT. SPACING, MASS, K FOR X & Y, K FOR XY:"
120 INPUT A,M,K1,K2
130 LET Q0=3.14159/A   Edge of Brillouin zone
140 LET W9=4*SQR(K1*K2)/M       } Constants in
150 LET R=SQR(K1/K2)            } equation for ω²
160 LET I8=0
170 PRINT "QX","QY","OMEGA"
180 FOR Q1=0 TO Q0 STEP Q0/10    Step out q_x
190 LET I8=I8+1
200 LET S=SIN(Q1*A/2)   } Parameters in
210 LET P=S*S           } equation for ω²
220 LET R1=1-P          }
230 LET W0=0
240 LET I9=0
250 FOR Q2=0 TO Q0 STEP Q0/10   Step up q_y
260 LET I9=I9+1
270 LET S=SIN(Q2*A/2)
280 LET Q=S*S
290 LET S1=1-Q
300 LET S9=R*R*(P-Q)*(P-Q)+(16/(R*R))*P*R1*Q*S1
310 LET W2=W9*(R*(P+Q)+(2/R)*(P*S1+Q*R1)-SQR(S9))     ω²
320 LET W=SQR(W2)   ω
330 LET W(I9,I8)=W   Store ω in grid
340 PRINT Q1,Q2,W
350 IF W<W0 THEN 370    } Find
360 LET W0=W            } maximum ω
370 NEXT Q2
380 PRINT
390 NEXT Q1
400 LET D0=0
410 FOR I=1 TO I9                        }
420 FOR J=1 TO I8                        } Count states
430 LET I0=INT(25*W(I,J)/W0+.5)+1        } into density
440 LET D(I0)=D(I0)+1  ,Density of states } of states vector
450 NEXT J                               }
460 NEXT I                               }
470 PRINT "OMEGA","# @ OMEGA"  }
480 FOR I=1 TO 26              }
490 LET W=(I-1)*W0/25          } Print density
500 PRINT W,D(I)               } of states
510 NEXT I                     }
520 END
```

in this array and solve the system of (coupled) equations analytically. The result for the angular frequency ω as a function of $q = (q_x, q_y)$ is

$$\omega^2 = \omega_0^2 \left\{ \sqrt{\frac{k_1}{k_2}}\,[P+Q] + 2\sqrt{\frac{k_2}{k_1}}\,(PS+QR) \pm \sqrt{\frac{k_1}{k_2}(P-Q)^2 + \frac{16k_2}{k_1}\,PRQS} \right\}$$

where

$$P = \sin^2\left(\frac{q_x a}{2}\right), \quad Q = \sin^2\left(\frac{q_y a}{2}\right), \quad R = 1 - P, \quad S = 1 - Q, \quad \text{and} \quad \omega_0^2 = \frac{4k_1k_2}{m}$$

Program 14.3(b)

```
OMOFQFOR      ω(q) for 2D square lattice

              REAL M,KX,KY
              DIMENSION OMEG(11,11),DENS(26)
              DATA A,M,KX,KY/3.14159,1.,1.,.1/    Lattice spacing, mass, spring constants
              Q0=3.14159/A    Brillouin zone edge
              OME=4.*SQRT(KX*KY)/M    } Parameters in
              R=SQRT(KX/KY)           } equation for ω²
              DO 40 I=1,11    Step out qx
              QX=(I-1)*Q0/10.
              S=SIN(QX*A/2.)  } Parameters in
              P=S*S           } equation for ω²
              R1=1.-P         }
              OMMAX=0.
              DO 40 J=1,11    Step up qy
              QY=(J-1)*Q0/10.
              S=SIN(QY*A/2.)
              Q=S*S
              S1=1.-Q
              S9=R*R*(P-Q)*(P-Q)+(16./(R*R))*P*R1*Q*S1
              OMEG2=OME*(R*(P+Q)+(2./R)*(P*S1+Q*R1)-SQRT(S9))    ω²
              OMEGA=SQRT(OMEG2)    ω
              OMEG(I,J)=OMEGA    Store ω in grid
              PRINT 100,QX,QY,OMEGA
   40         OMMAX=AMAX1(OMMAX,OMEGA)    Find maximum ω
              D0=0.
              DO 45 I=1,11                          } Count states
              DO 45 J=1,11                          } into density of
              I0=INT(25.*OMEG(I,J)/OMMAX+.5)+1      } states vector
   45         DENS(I0)=DENS(I0)+1.                  }
              DO 50 I=1,26                 } Print density
              OMEGA=(I-1)*OMMAX/25.        } of states vector
   50         PRINT 101,OMEGA,DENS(I)      }
              STOP
  100         FORMAT(1X,3F10.4)
  101         FORMAT(1X,2F10.4)
              END
```

The computer is used to display this result. You can see some of the behavior of the analytic solution, but the details of the solution need a calculational machine. One program to do the necessary calculations is OMOFQ [Programs 14.3(a), (b)]. One way to display the result is to use terminal (or printer) plotting where the character printed at each position is given by the value of ω and the position is given by (q_x, q_y). The accompanying Figure 14.6 shows such a plot.

14.5 Two-Dimensional Nearly Free Electrons

The final example of uses of the computer in introductory solid-state physics deals with nearly free electrons in two (or more) dimensions. In one dimension, standard texts show derivations and results of the energy E versus wave vector k. By allowing electrons near Brillouin-zone edges to interact (via Bragg reflection) with the nearly

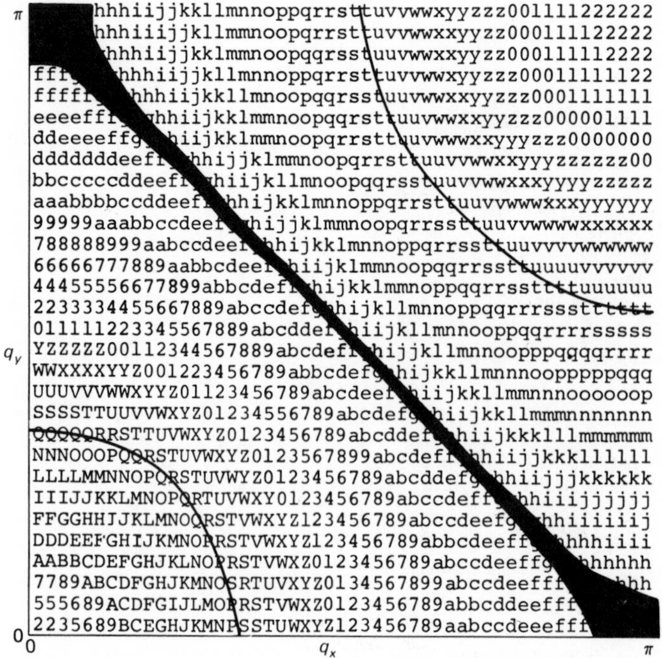

Figure 14.6 ω-versus-q plot for a two-dimensional square lattice with nearest-neighbor (spring constant 1) and next-nearest-neighbor (spring constant 0.5) spring constants. One-quarter of the first Brillouin zone is shown with the origin at the lower left. The characters represent the magnitude of ω at each point (q_x, q_y). Several constant ω contours are shown.

energy degenerate electrons a reciprocal lattice vector away, energy gaps appear in the E-versus-k spectrum. Second-order degenerate perturbation theory is used to find the correct energy E for a state in terms of the matrix element M for the reciprocal lattice interaction and the unperturbed energies E_1 and E_2 of the two interacting states. $E = \frac{1}{2}\ [(E_1 + E_2) \pm \sqrt{(E_1 - E_2)^2 + 4M^2}]$ (which comes directly from diagonalizing the 2×2 perturbation matrix).

Before moving to full band structure calculations, it is helpful to see the nearly free electron calculation in two dimensions. In two dimensions you have energy gaps (different energies on adjacent points across the Brillouin-zone face) and wave vector gaps (a change in k across a zone face when following the same energy contour). These two-dimensional effects are better illustrations of the meanings of the terms holes, magnetic breakdown, multiple-piece Fermi surfaces, etc., than are the results in one dimension.

The one-dimensional result is good near zone boundaries and not so good far from zone boundaries. The same will be true of our two-dimensional nearly free electron approximation. In two dimensions near the corners of Brillouin zones, more than two states separated by reciprocal lattice vectors may have nearly degenerate energies; at such points it is appropriate to diagonalize a larger (3×3 or 4×4, etc.) perturbation matrix.

The computer is used in a straightforward way. You step along through **k** space; at each **k** you calculate the energies of all the states related to **k** by reciprocal lattice vectors, and you choose the most nearly degenerate such state. You then diagonalize the 2×2 matrix for these two states mixed by the appropriate matrix element M. Having calculated the energy (and printed it or plotted it at the terminal), you move on to the next **k**. The result is the nearly free electron energies for the system.

Program 14.4(a)

```
NEARFRBA    2D nearly free electrons

100 DIM D(26)
110 PRINT "MAX. K";
120 INPUT K0
130 LET E0=2*K0*K0    Energy for maximum wave vector
140 FOR I=1 TO 8                                           } Reciprocal
150 READ K(I,1),K(I,2),M(I)                                } lattice vectors
160 NEXT I                                                 } and matrix
170 DATA 0,2,.2, 2,0,.2, 2,2,.2                            } elements
180 DATA 0,4,.2, 2,4,.2, 4,4,.2, 4,2,.2, 4,0,.2            }
190 PRINT "KX","KY","ENERGY"
200 LET D0=0
210 FOR J=0 TO 10    Step through ky
220 LET K2=J*K0/10    ky
230 FOR I=0 TO 10    Step through kx
240 LET K1=I*K0/10    kx
250 LET E1=K1*K1+K2*K2    Energy at (kx, ky) [free electron]
260 LET E8=9999                                                      } Find most
270 FOR I0=1 TO 8                                                    } nearly
280 LET E9=(K1-K(I0,1))*(K1-K(I0,1))+(K2-K(I0,2))*(K2-K(I0,2))       } degenerate
290 LET E7=ABS(E1-E9)                                                } energy and
300 IF E7>E8 THEN 340                                                } corresponding
310 LET E2=E9                                                        } reciprocal
320 LET E8=E7                                                        } lattice vector
330 LET M=M(I0)                                                      }
340 NEXT I0                                                          }
350 IF E1<>E2 THEN 370
360 LET E2=E1+.00001
370 LET E=(E1+E2)/2+SGN(E1-E2)*SQR((E1-E2)*(E1-E2)/4+M*M)    Nearly free energy
380 PRINT K1,K2,E
390 LET I0=INT(25*E/E0+.5)+1    } Store state in
400 LET D(I0)=D(I0)+1           } density of states vector
410 NEXT I
420 NEXT J
430 PRINT "ENERGY","# WITH ENERGY"    } Print
440 FOR I=1 TO 26                     } density of
450 LET E=(I-1)*E0/25                 } states
460 PRINT E,D(I)                      }
470 NEXT I                            }
480 END
```

Program 14.4(b)

```
NEARFRFO        2D nearly free electrons

           DIMENSION DENS(26)
           REAL KMAX,KX,KY,K(8,2),MATEL(8)
           DATA K(1,1),K(1,2),MATEL(1)/0.,2.,.2/   )
           DATA K(2,1),K(2,2),MATEL(2)/2.,0.,.2/   | Reciprocal
           DATA K(3,1),K(3,2),MATEL(3)/2.,2.,.2/   | lattice points
           DATA K(4,1),K(4,2),MATEL(4)/0.,4.,.2/   > and
           DATA K(5,1),K(5,2),MATEL(5)/2.,4.,.2/   | matrix
           DATA K(6,1),K(6,2),MATEL(6)/4.,4.,.2/   | elements
           DATA K(7,1),K(7,2),MATEL(7)/4.,2.,.2/   |
           DATA K(8,1),K(8,2),MATEL(8)/4.,0.,.2/   )
           KMAX=1.
           E0=2.*KMAX*KMAX   Energy for maximum wave vector
           D0=0.
           DO 50 J=1,11   Step through ky
           KY=(J-1)*KMAX/10.
           DO 50 I=1,11   Step through kx
           KX=(I-1)*KMAX/10.
           E1=KX*KX+KY*KY   Energy at (kx, ky) [free electron]
           E8=9999.                                  )
           DO 40 I0=1,8                              |
           ERLP=(KX-K(I0,1))**2+(KY-K(I0,2))**2      | Find most
           DELE=ABS(E1-ERLP)                         | nearly degenerate
           IF (DELE-E8) 35,35,40                     > energy and
35         E2=ERLP                                   | corresponding
           E8=DELE                                   | reciprocal lattice vector
           AM=MATEL(I0)                              |
40         CONTINUE                                  )
           IF (E1-E2) 47,45,47
45         E2=E1+.00001
47         E=(E1+E2)/2.+SIGN(E1-E2)*SQRT((E1-E2)**2/4.+AM**2)   Nearly free energy
           PRINT 100,KX,KY,E
           I0=INT(25.*E/E0+.5)+1     ) Store state in
50         DENS(I0)=DENS(I0)+1.      ) density of states vector
           DO 60 I=1,26              )
           E=(I-1)*E0/25.            > Print density
60         PRINT 101,E,DENS(I)       ) of states vector
           STOP
100        FORMAT(1X,3E12.4)
101        FORMAT(1X,2E12.4)
           END
```

One program which does these calculations is NEARFR [Programs 14.4(a), (b)]. Figures 14.7(a) and (b) are terminal plots where the character printed represents the energy E and the position on the plot is (k_x, k_y).

This chapter has introduced four ways the computer can be used to deepen your understanding of basic phenomena in introductory solid-state physics. There are many other applications (calculations of specific heats from measured ωq relations for solids; simulations of x-ray, neutron, and electron-diffraction experiments; calculations of current-voltage tunneling characteristics; calculations of the approximate band structure for amorphous solids, etc.). This chapter has only scratched the surface; now you can find additional interesting applications on your own.

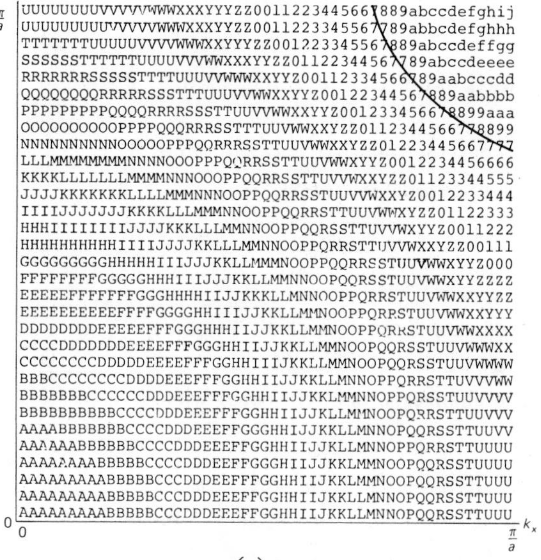

(a)

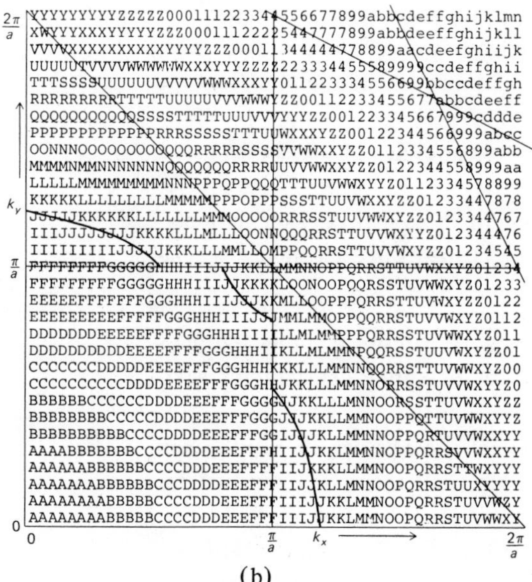

(b)

Figure 14.7 Two-dimensional nearly free electron calculations. The matrix elements connecting states are all 10% of the energy at the zone edge. (a) One quarter of the first Brillouin zone is shown. The bending of the higher energy contours near the zone corner is apparent. (b) Pieces of the first, second, third, fourth, fifth, and sixth zones. The ΔE and ΔK gaps near each zone face are clear.

Problems

The following illustrate some ways the material discussed in this chapter can be used.

1. a. Draw the Brillouin zone and Harrison construction (for two and three electrons per cell) for the following two-dimensional lattices:
   (1) Hexagonal
   (2) Oblique
   (3) Centered rectangular
   b. Map the pieces of the Fermi surfaces (for two and three electrons per cell) back into the first zone and identify electron and hole surfaces.
   c. Sketch on your pictures from part b the changes which will occur when the nearly free electron model is used.

2. Consider the square phonon pulse in one dimension. Using the ideas of Chapter 10 and the phonon velocities from the one-dimensional chain, propagate the pulse down the chain. Use pulses centered near $q = 0$ in the continuum region and near the Brillouin-zone edge.

3. Consider a two-dimensional nearly free electron model for a square lattice.
   a. Using matrix elements only for the edges of the first Brillouin zone, demonstrate the rounding of energy contours and the growth of zone edge gaps as the matrix element grows.
   b. Now include matrix elements for higher zone boundaries. Examine part a again.
   c. Near the corner of the first zone, effects of three matrix elements apply. Consider how to include all three matrix elements in each energy calculation.

4. a. Calculate the ω-versus-q contours for a two-dimensional spring and mass square lattice. Consider several values of the spring constant for next-nearest neighbors.
   b. Discuss the behavior of the lattice as the next-nearest-neighbor spring constant starts at 0 and rises to become equal to the spring constant for nearest neighbors.
   c. For a realistic interatomic potential and a solid of your choice, estimate relative values of the nearest- and next-nearest-neighbor spring constants.

appendix: some numerical methods

A.1 Introduction

This appendix is a brief introduction to numerical methods, that is, the ways that you can solve real problems using computers. There are a number of good books in the field, including B. Carnahan, H. A. Luther, and J. O. Wilkes, *Applied Numerical Methods,* John Wiley & Sons, New York, 1969; P. Henrici, *Discrete Variable Methods in Ordinary Differential Equations,* John Wiley & Sons, New York, 1962; and B. W. Arden, *Introduction to Digital Computing,* Addison-Wesley Publishing Company, Reading, Mass. 1963. The Henrici book deals more directly with the numerical problems that you have been studying (solutions to differential equations). Carnahan, Luther, and Wilkes does a commendable job on differential equations as well as on other problems.

A.2 Ordinary Differential Equations

The basic problem in numerical methods for differential equations is the following. How do you accurately express a problem involving continuous variables (a differential equation) in terms of discrete variables and finite-step mathematical methods (a finite difference equation)?

There turn out to be several major questions that need to be answered.

1. Does the numerical procedure you choose *represent* your problem? The answer is practically always yes, but you must be sure.

2. Does the numerical procedure *converge* to the true solution of the differential equation? In other words, if you take infinitesimal steps, will you actually get the continuous variable solution? Again, the answer is usually yes, but even so, convergence is important. You will usually want to get close to the true solution with as large a step as possible. The more highly convergent your method, the larger step size you can use.

3. Is the solution to the problem *stable?* If you make a small error at any point in the calculation, does the error get worse and worse? If your problem is unstable, solutions finally blow up.

4. Is there an *optimal step size* that minimizes truncation error (due to the accuracy of the method) and round-off error (due to the accuracy of the computer)?

Any nth-order differential equation can be written as n first-order differential equations. For example, the second-order differential equation $d^2y/dt^2 = F/m$ can be written as two first-order differential equations: $dv/dt = F/m$ and $dy/dt = v$. We will limit the discussion to the solution of first-order differential equations. (There are ways to handle higher-order differential equations directly, but you need not deal with them right now.)

The problem, then, is to solve the first-order ordinary differential equation of the form

$$\frac{dy}{dx} = f(x, y)$$

on the interval $a < x < b$ with an initial condition $y_0 = y(a)$. (x is any independent variable; in many of your problems the independent variable has been time.) In general, you can not find a solution to this problem in analytic form. Most first-order differential equations do not have solutions that you can write down as an equation. So if analytic techniques fail, you must try numerical techniques.

Numerical solutions begin by dividing the (closed) interval $a < x < b$ into n pieces, $\Delta x = (b - a)/n$, and labeling the coordinates of the end points of each interval $x_{i+1} = x_i + \Delta x$ (and $x_0 = a$). Since you are using numerical methods, the solution to your problem is in the form of a table (see Table A.1). Your solution is the first two columns; the last two columns are unknown. $y(x)$ is the (unknown) complete solution to your differential equation. Because you use only finite steps in your solution (instead of infinitesimal steps), you will introduce some error. This error is called a

Table A.1

	Numerical solution	True solution	Truncation error ϵ
x_0	Y_0	$y(x_0)$	0 (initial conditions)
x_1	Y_1	$y(x_1)$	$y(x_1) - Y_1$
x_2	Y_2	$y(x_2)$	$y(x_2) - Y_2$
.	.	.	.
.	.	.	.
x_i	Y_i	$y(x_i)$	$y(x_i) - Y_i$
x_{i+1}	Y_{i+1}	$y(x_{i+1})$	$y(x_{i+1}) - Y_{i+1}$
.	.	.	.

"truncation" or "discretization" error. (For steps which are too small, round-off error due to the finite storage accuracy of the computer dominates.) You must try to minimize that error. In order to do so, you will probably end up using a numerical method which you have studied. There are several common techniques.

There are two kinds of numerical methods of solutions to first-order differential equations: one-step methods (where Y_{i+1} is given in terms of information about values of the function Y_i and its derivative in the interval $x_i \leqslant x \leqslant x_{i+1}$) and multistep methods [where you use information outside the $(i+1)$st interval]. Almost all the solutions you have used so far are of the one-step kind.

If you look at the problem graphically, the reason for the difficulty is clearer. Consider a graph of the true solution $y(x)$ and several of your discrete lattice points $[x_i]$, Figure A.1. Suppose you found Y_i exactly, that is, suppose $Y_i = y(x_i)$. You want to find $y(x_{i+1})$ exactly [that is, you want $Y_{i+1} = y(x_{i+1})$], but all you have is some information about derivatives of $y(x)$ at the discrete points, x_i. The simplest approximation is to say $Y_{i+1} = Y_i + \Delta x(dy/dx)|_{x_i}$, where $dy/dx|_{x_i}$ is the derivative of $y(x)$ evaluated at x_i. You have approximated $y(x_{i+1})$ by moving along the tangent to $y(x)$ at the point x_i. The approximation is called Euler's method, which is useful because it is so simple. As you can see from the figure, you may well miss the curve at x_{i+1}.

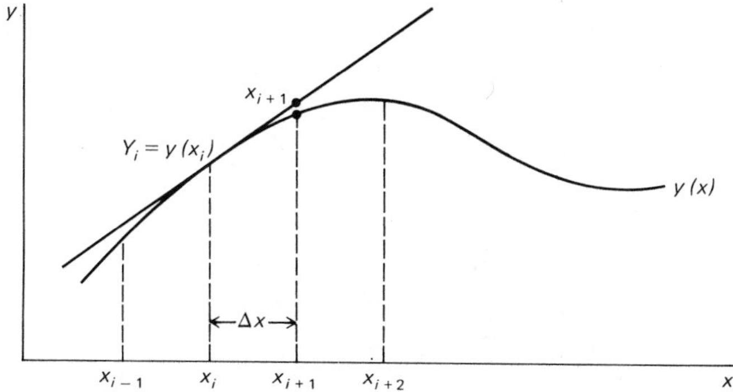

Figure A.1 A graphical interpretation of Euler's method of numerical analysis.

By the mean value theorem, there is some point, in $x_i \leqslant x \leqslant x_{i+1}$, for which the tangent to the curve $y(x)$ is parallel to the slope of the line between $(x_{i+1}, y(x_{i+1}))$ and $(x_i, Y_i = y(x_i))$. The one-step methods all try to find that point or that correct slope in some appropriate way. Euler's method uses the slope at one end of the interval. The half-step methods you have been using approximate the correct slope by the slope at the midpoint of the interval–in general, a better choice.

There are a whole series of methods (called Runge-Kutta methods) which use weighted averages of slopes in the interval to get an increasingly better approximation to the correct slope. The half-step method is (essentially) a second-order Runge-Kutta method. [Second order in this context means that if you expand the true solution around x_i as a Taylor series, the method will be exactly right through the Taylor series term in $(\Delta x)^2$. If you are interested in proving that any particular method is second (or third or fourth) order, see Carnahan, Luther, and Wilkes.] Research has shown that predictor-corrector methods are sometimes more accurate for the same computer time than are Runge-Kutta methods.

Multipoint methods use a somewhat different approach. A number of such methods use Y_{i-1}, Y_{i-2}, etc., as well as Y_i and $dy/dx|_{x_i}$ in order to fit a polynomial to a curve and extrapolate to Y_{i+1}. These methods can be useful when only limited information is available in the interval $x_i \leqslant x \leqslant x_{i+1}$. The methods are usually not very easy to use to start a calculation, for which you need some other method.

Sometimes you can guess a more accurate (higher-order) numerical method. For example, in approximating the correct slope by (1) averaging the slopes at (x_i, Y_i) and at $(x_{i+1}, \widetilde{Y}_{i+1})$ where $\widetilde{Y}_{i+1}$ is an estimated next Y value, or (2) averaging those two slopes with the slope at the midpoint, you are using a second-order Runge-Kutta method in the first instance and a third-order method in the second.

By using Taylor expansions, you can often prove that Euler or Runge-Kutta methods do (1) represent your differential equation correctly and (2) converge to the true solution so that

$$\lim_{\Delta x \to 0} Y_i = y(x_i)$$

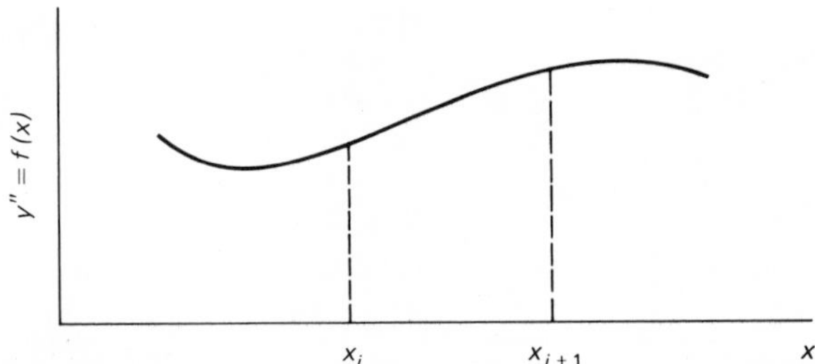

Figure A.2 A general functional dependence of a second derivative.

There are pathological cases in which, even though your solution is representative and convergent, it is still not stable. Stability proofs are usually very complicated. The best way you can test stability is to see if the solution blows up. If the solution blows up, you are better off finding a new procedure. Instability is sometimes a property of the differential equation and the particular boundary condition rather than of your choice of algorithm.

A.3 A General Higher-Order Method

There is one class of higher-order convergence methods which is easy to explain and easy to apply, especially in cases where some derivative is a function only of one variable and not its derivatives. This class of methods is similar to multipoint fitting methods but uses information only from the present interval. The method has the advantage of being fast, of being applicable to any order of differential equation, and of showing increasing accuracy in a manner disconnected with step size. The method will be illustrated for a second-order differential equation of the form $y'' = f(x)$.

Suppose that the second derivative y'' has a curve such as is shown in Figure A.2 and that you have found the solution at x_i. The problem is to find the solution at $x_{i+1}, y(x_{i+1})$, and possibly the derivative $y'(x_{i+1})$. The idea is to fit a polynomial to the second derivative over the (closed) interval $[x_i, x_{i+1}]$ and then integrate the polynomial. The integration can be performed analytically and written as a single line in the program.

First, find the coefficients in the equation

$$y''(x_i + \delta x) = A_1 + A_2 \ \delta x + A_3(\delta x)^2 \cdots A_{n+1}(\delta x)^n$$

for $0 < \delta x < \Delta x$. The solution is then

$$Y_{i+1} = Y_i + Y_i' \Delta x + A_1 \frac{(\Delta x)^2}{2} + A_2 \frac{(\Delta x)^3}{6} \cdots A_{n+1} \frac{(\Delta x)^{n+2}}{(n+1)(n+2)}$$

The first derivative is

$$Y_{i+1}' = Y_i' + A_i \ \Delta x + A_2 \frac{(\Delta x)^2}{2} \cdots A_{n+1} \frac{(\Delta x)^{n+1}}{n+1}$$

Clearly, the method extrapolates to any order of differential equation. The solution is an $(n + 2)$ Runge-Kutta method; it is also very similar to Simpson's rule integration if $n = 2$ (quadratic fit).

The polynomial fit in the interval is easy. For simplicity, choose the end points and equally spaced interior points. For the cases $n = 2$ and $n = 3$, the coefficients can be written down algebraically. For larger n, matrix multiplication is easier.

To find the coefficients A_i, you need to solve the $(n + 1)$ equations (where $n = \Delta x/k - 1$ and $\eta = \Delta x/k$, for $k = 3$):

$$\begin{aligned} y''(x_i) &= A_1 \\ y''(x_i + \eta) &= A_1 + A_2\eta + A_3\eta^2 \cdots A_{n+1}(\eta)^n \\ y''(x_i + 2\eta) &= A_1 + A_2(2\eta) + A_3(2\eta)^2 \cdots A_{n+1}(2\eta)^n \\ y''(x_{i+1}) &= A_1 + A_2(\Delta x) + A_3(\Delta x)^2 \cdots A_{n+1}(\Delta x)^n \end{aligned}$$

Call the left side the matrix $F(\,)$, and set up the square matrix M:

$$M = \begin{Bmatrix} 1 & 0 & 0 & \cdots & \cdots & \cdots & 0 \\ 1 & \eta & \eta^2 & \cdots & \cdots & \cdots & \eta^n \\ 1 & (2\eta) & (2\eta) & \cdots & \cdots & \cdots & (2\eta)^n \\ \cdots & \cdots & \cdots & \cdots & \cdots & \cdots & \cdots \\ 1 & \Delta x & \Delta x & \cdots & \cdots & \cdots & (\Delta x)^n \end{Bmatrix}$$

The coefficient matrix A is then $A = M^{-1}F$.

To increase convergence in this method, you can often either increase the order of the fitted polynomial n or decrease the step size Δx. One is sometimes easier than the other in a particular problem.

Often in physics you are integrating an equation in time but know the second (or first) derivative in space (for example, $\mathbf{F} = m\mathbf{a}$, where the force is a known function of position). You can thus get an approximate set of values of $y''(t_i + \delta t)$ by assuming constant acceleration through the interval so that

$$y''(t_i + \delta t) = y''(R_i + y'(R_i)\,[\delta t] + y''(R_i)\,[\tfrac{1}{2}\,(\delta t)^2\,])$$

This convergence method has worked well even in cases where the derivative whose values you know varies very quickly in some places and very slowly in others. Examples include $\mathbf{F} = m\mathbf{a}$ calculations in a 6–12 potential, Schrödinger equation problems, and field-line problems where the lines or the equipotential contours bend sharply. The greatest advantage of this method is probably that, even though it is a high-order Runge-Kutta method, it is easy to explain and understand.

A.4 Faster Calculations of Sine, Cosine, and Square Root

On many computers the evaluation of supplied (library) functions is relatively slow. For example, a single evaluation of the sine of an angle may take 400 to 500 ms on a machine that multiplies in about 5 μs. A number of programs in this book use the sine, cosine, or square-root functions repeatedly (often in interactive calculations); such programs use a lot of machine time. There are ways to calculate values of these functions more quickly on a typical machine; the ideas can also be extended to other functions.

There are several ways to calculate fast sines and cosines. In some situations, such as Fourier series calculations, you only use angles that are at equal angle steps. In such cases, it is fastest to use the sum of angle formulas:

$$\sin(\theta + \Delta\theta) = \sin(\theta)\cos(\Delta\theta) + \cos(\theta)\sin(\Delta\theta)$$
$$\cos(\theta + \Delta\theta) = \cos(\theta)\cos(\Delta\theta) - \sin(\theta)\sin(\Delta\theta)$$

and step through the angles. By evaluating $\sin(\Delta\theta)$ and $\cos(\Delta\theta)$ once outside the loop, each evaluation of sine and cosine becomes four multiplications and two additions. On most computers such a scheme is much faster than the evaluation of supplied (library) sine and cosine functions. A program illustrating this method is given in Programs A.1(a) and (b).

Program A.1(a)

```
FSIN1BA

100 LET T1=0
110 LET T2=3.14159
120 LET D=(T2-T1)/20     Δθ
130 LET S1=SIN(D)     sin(Δθ)
140 LET C1=COS(D)     cos(Δθ)
150 LET S=SIN(T1-D)   } Initial sine and
160 LET C=COS(T1-D)   } cosine (one step back)
170 FOR T=T1 TO T2 STEP D
180 LET S0=S*C1+C*S1  }
190 LET C=C*C1-S*S1   } Sum formulas
200 LET S=S0          }
210 PRINT 180*T/3.14159,C,S
220 NEXT T
230 END
```

Program A.1(b)

```
FSIN1FO

          THET1=0.
          THET2=3.14159
          DTHET=(THET2-THET1)/20.     Δθ
          SDEL=SIN(DTHET)    sin(Δθ)
          CDEL=COS(DTHET)    cos(Δθ)
          FSIN=SIN(THET1-DTHET)   } Initial sine and
          FCOS=COS(THET1-DTHET)   } cosine (one step back)
          DO 10 I=1,21
          AI=I-1
          THET=THET1+AI*DTHET
          FSTEM=FSIN*CDEL+FCOS*SDEL  }
          FCOS=FCOS*CDEL-FSIN*SDEL   } Sum formulas
          FSIN=FSTEM                 }
10        PRINT 101,THET,FCOS,FSIN
          STOP
101       FORMAT(1X,3E10.4)
          END
```

In most situations, the angles you need do not come in equal steps. One strategy for fast sines and cosines involves storing some values (say at every 10° interval from 0° through 360°) and then using the sum formula and the stored values to calculate sines and cosines. You store values of the sine and cosine in vectors [say *C*( ) and *S*( )] at the start of the program. When you need a sine or cosine, you then: (1) find the closest angle entry in the vectors, (2) find the difference angle $\Delta\theta$ between your angle and the stored angle, (3) use the expansion formulas for $\sin(\Delta\theta)$ and $\cos(\Delta\theta)$ [$\sin(\Delta\theta) \approx \Delta\theta - \frac{1}{6}(\Delta\theta)^3$; $\cos(\Delta\theta) \approx 1 - \frac{1}{2}(\Delta\theta)^2$], and finally (4) use the sum formula to find the sine and cosine of your angle. A program illustrating this method is shown in Programs A.2(a) and (b).

Program A.2(a)

```
FSIN2BA

100 DIM S(37),C(37)
110 FOR I=1 TO 37                          }
120 LET S(I)=SIN((I-1)*3.14159/18)         } Store sines and cosines
130 LET C(I)=COS((I-1)*3.14159/18)         }
140 NEXT I                                 }
150 PRINT "ANGLE(DEGS.)?"
160 INPUT T
170 LET T=3.14159*T/180
180 LET T1=T-6.28318*INT(T/6.28318)   Reduce angle to [0, 2π]
190 LET N1=1+INT(T1/(3.14159/18)+.5)   Entry in vectors
200 LET D1=T1-(N1-1)*3.14159/18   Δθ
210 LET S1=D1-D1*D1*D1/6   sin(Δθ)
220 LET C1=1-D1*D1/2   cos(Δθ)
230 LET S=S(N1)*C1+C(N1)*S1   sine
240 LET C=C(N1)*C1-S(N1)*S1   cosine
250 PRINT S,C
260 GOTO 150
270 END
```

Program A.2(b)

```
FSIN2FO

          DIMENSION S(37),C(37)
          DO 10 I=1,37                     }
          AI=I-1                           } Store sines
          S(I)=SIN(AI*3.14159/18.)         } and cosines
10        C(I)=COS(AI*3.14159/18.)         }
20        READ(5,100)THETO
          IF (THETO-999.) 30,60,30
30        THET=3.14159*THETO/180.
          THET1=THET-6.28318*AINT(THET/6.28318)   Reduce angle to [0, 2π]
          N=1+INT(THET/(3.14159/18.)+.5)   Entry in vectors
          AN=N-1
          DELT=THET1-AN*3.14159/18.   Δθ
          SD=DELT-DELT**3/6.   sin(Δθ)
          CD=1.-DELT**2/2.   cos(Δθ)
          SI=S(N)*CD+C(N)*SD   sine
          CO=C(N)*CD-S(N)*SD   cosine
          PRINT 101,THETO,SI,CO
          GOTO 20
60        STOP
100       FORMAT(F10.5)
101       FORMAT(1X,3E10.4)
          END
```

If some definite accuracy (say 1° accuracy) is sufficient in your calculations, the fastest form of sine and cosine evaluation often is storing sines and cosines at those (1°) intervals and then pulling out the right elements when you need them.

It is common to need a series of square roots of numbers fairly close together. Examples are the calculations of distances in algorithmic calculations of electrostatic field lines.

One way to perform fast calculations of square roots in such situations is based on Newton's formula for the square root. If S is an approximation to the square root of N, then a better approximation, S', is given by $S' = \frac{1}{2}(S + N/S)$.

In trajectory or field-line calculations, each value of the distance is a good approximation to the next value of distance. At most, two applications of Newton's formula usually give sufficient accuracy at each new step. [Because the errors are cumulative, you might want to call the supplied square root function every so often (say after every 20 uses of Newton's formula)]. You always start the procedure by calling the supplied square root for the first distance on a line. A program illustrating this fast square-root procedure is as given in Programs A.3(a) and (b).

Numerical methods and their applications are a growing field of applied mathematics. Numerical methods are rapidly becoming important in almost every field of human endeavor, from the classics (where computers are used to analyze literature and establish authorship) to the physical sciences. Much of the progress in numerical methods is fairly recent, so there is still a lot to be done.

Program A.3(a)

```
FSQRBA

100 FOR Y=1 TO 4
110 LET R=SQR(Y*Y+1)
120 FOR X=1 TO 4 STEP .2
130 LET R2=X*X+Y*Y      r2
140 LET R=(R+R2/R)/2    } Two uses of
150 LET R=(R+R2/R)/2    } Newton's formula
160 PRINT X,Y,R
170 NEXT X
180 NEXT Y
190 END
```

Program A.3(b)

```
FSQRFO

          DO 10 I=1,4
          Y=I
          R=SQRT(Y*Y+1.)
          DO 10 J=1,31
          AJ=J-1
          X=1.+AJ*.1
          RSQ=X*X+Y*Y    r2
          R=(R+RSQ/R)/2.     } Two uses of
          R=(R+RSQ/R)/2.     } Newton's formula
10        PRINT 100,X,Y,R
          STOP
100       FORMAT(1X,3F10.4)
          END
```

index

ABCDEFGHIJ–A–79876